Springer Tracts in Modern Physics 92

Springer Tracts in Modern Physics

66* **Quantum Statistics in Optics and Solid-State Physics**
With contributions by R. Graham, F. Haake

67* **Conformal Algebra in Space-Time and Operator Product Expansion**
By S. Ferrara, R. Gatto, A. F. Grillo

68* **Solid-State Physics** With contributions by D. Bäuerle, J. Behringer, D. Schmid

69* **Astrophysics** With contributions by G. Börner, J. Stewart, M. Walker

70* **Quantum Statistical Theories of Spontaneous Emission and their Relation to Other Approaches** By G. S. Agarwal

71 **Nuclear Physics** With contributions by J. S. Levinger, P. Singer, H. Überall

72 **Van der Waals Attraction:** Theory of Van der Waals Attraction By D. Langbein

73 **Excitons at High Density** Edited by H. Haken, S. Nikitine. With contributions by V. S. Bagaev, J. Biellmann, A. Bivas, J. Goll, M. Grosmann, J. B. Grun, H. Haken, E. Hanamura, R. Levy, H. Mahr, S. Nikitine, B. V. Novikov, E. I. Rashba, T. M. Rice, A. A. Rogachev, A. Schenzle, K. L. Shaklee

74 **Solid-State Physics** With contributions by G. Bauer, G. Borstel, H. J. Falge, A. Otto

75 **Light Scattering by Phonon-Polaritons** By R. Claus, L. Merten, J. Brandmüller

76 **Irreversible Properties of Type II Superconductors** By. H. Ullmaier

77 **Surface Physics** With contributions by K. Müller, P. Wißmann

78 **Solid-State Physics** With contributions by R. Dornhaus, G. Nimtz, W. Richter

79 **Elementary Particle Physics** With contributions by E. Paul, H. Rollnick, P. Stichel

80* **Neutron Physics** With contributions by L. Koester, A. Steyerl

81 **Point Defects in Metals I:** Introduction to the Theory 2nd Printing
By G. Leibfried, N. Breuer

82 **Electronic Structure of Noble Metals, and Polariton-Mediated Light Scattering**
With contributions by B. Bendow, B. Lengeler

83 **Electroproduction at Low Energy and Hadron Form Factors**
By E. Amaldi, S. P. Fubini, G. Furlan

84 **Collective Ion Acceleration** With contributions by C. L. Olson, U. Schumacher

85 **Solid Surface Physics** With contributions by J. Hölzl, F. K. Schulte, H. Wagner

86 **Electron-Positron Interactions** By B. H. Wiik, G. Wolf

87 **Point Defects in Metals II:** Dynamical Properties and Diffusion Controlled Reactions
With contributions by P. H. Dederichs, K. Schroeder, R. Zeller

88 **Excitation of Plasmons and Interband Transitions by Electrons** By H. Raether

89 Giant Resonance Phenomena in **Intermediate-Energy Nuclear Reactions**
By F. Cannata, H. Überall

90* **Jets of Hadrons** By W. Hofmann

91 **Structural Studies of Surfaces**
With contributions by K. Heinz, K. Müller, T. Engel, K. H. Rieder

92 **Single-Particle Rotations in Molecular Crystals** By W. Press

93 **Coherent Inelastic Neutron Scattering in Lattice Dynamics** By B. Dorner

* denotes a volume which contains a Classified Index starting from Volume 36.

Werner Press

Single-Particle Rotations in Molecular Crystals

With 53 Figures

Springer-Verlag
Berlin Heidelberg New York 1981

Dr. Werner Press

Kernforschungsanlage Jülich, Institut für Festkörperforschung,
D-5170 Jülich, Fed. Rep. of Germany

Manuscripts for publication should be addressed to:

Gerhard Höhler

Institut für Theoretische Kernphysik der Universität Karlsruhe
Postfach 6380, D-7500 Karlsruhe 1, Fed. Rep. of Germany

Proofs and all correspondence concerning papers in the process of publication should be addressed to:

Ernst A. Niekisch

Haubourdinstrasse 6, D-5170 Jülich 1, Fed. Rep. of Germany

ISBN 3-540-10897-1 Springer-Verlag Berlin Heidelberg New York
ISBN 0-387-10897-1 Springer-Verlag New York Heidelberg Berlin

Library of Congress Cataloging in Publication Data. Press, W. (Werner), 1942-. Single-particle rotations in molecular crystals. (Springer tracts in modern physics; 92). Bibliography: p. Includes index. 1. Molecular crystals. 2. Molecular rotation. 3. Neutrons-Scattering. I. Title. II. Series. QC1.S797 vol. 92 [QD921] 539s [548'.8] 81-9051 AACR2

Offset printing and bookbinding: Brühlsche Universitätsdruckerei, Giessen
2153/3130 – 5 4 3 2 1 0

Preface

Advances in experimental techniques always induce considerable scientific progress as well. This is particularly true for neutron scattering where the availability of new instruments (e.g. spectrometers for small-angle scattering, diffuse scattering and high-energy resolution) has stimulated a variety of fields in physics, chemistry and biology. One area, investigated by both physicists and chemists, is that of molecular crystals. Interesting questions concern molecular orientations in various crystal surroundings and the changes from a disordered to an ordered arrangement displayed by some crystals.

Here we mainly deal with a dynamical aspect, namely the rotational motion of single molecules. Its nature strongly depends on the interaction of a given molecule with its neighbours and - very importantly - on the temperature. In this book recent experimental results are described. They illustrate the importance of high-resolution neutron spectroscopy for a better understanding of molecular rotations. At the same time, recent theoretical models are reviewed which often have served as guidelines for new experiments. The book is intended to describe the state of the art, not only to neutron scatterers, but also to solid-state physicists and chemists interested in molecular systems.

Molecular crystals have been and continue to be an object of active research in the neutron scattering group in Jülich. The author thanks all his colleagues, especially H.H. Stiller, M. Prager, H. Grimm, K.D. Ehrhardt, and U. Buchenau, for a long and fruitful collaboration. On the way to a better understanding of molecular crystals many theoretical aspects had to be clarified and some still remain open. It is a particular pleasure to acknowledge the advice from A. Hüller as well as many stimulating discussions with him on numerous topics over a long period of time. The author is very grateful to A. Hüller, R. Mueller, P. Grosse, and U. Felderhof for important suggestions concerning the scientific content of the manuscript and its formulation.

Jülich, June 1981 *Werner Press*

Contents

1. Introduction

Starting with initial measurements in the 1920's—at this time CLUSIUS /1.1/ discovered specific-heat anomalies, e.g., in nitrogen—molecular crystals have attracted increasing attention. Most early experiments concentrated on macroscopic quantities such as specific-heat or dielectric constants. However, without microscopic probes the detailed mechanism of phase transitions and elementary excitations in molecular crystals remained a puzzle.

This has changed considerably in the last decade or so, when systematic efforts have been made to learn about molecular crystals both experimentally and theoretically. In the meantime numerous structures of molecular crystals have been solved, and orientational order-disorder transitions as well as rotational dynamics have been investigated in great detail. Some of this has been covered in two recent reviews 1) *The Plastically Crystalline State* edited by SHERWOOD /1.2/ and 2) *Disorder in Crystals* by PARSONAGE and STAVELEY /1.3/. (The latter discussed positional, magnetic and—mainly—orientational disorder.) Much of the recent development, however, is not contained or is reviewed only briefly.

This is particularly true for rotational dynamics, which is one of the most fascinating aspects of molecular solids. The field may be divided (somewhat artificially) into collective and single-particle rotational motions. In regard to collective rotational motions, one is concerned with phase relations between the rotational states of different molecules. These phase relations show up in a wave vector dependence of the rotational excitations. Their origin is the angle-dependent interaction between the molecules. The excitations, which are called librational or torsional modes, are analogous to the phonon modes describing the collective translational modes in a crystal. Single-molecule rotations, on the other hand, describe the rotational motion of a single molecule in its surroundings. The surroundings are approximated as an angle-dependent potential which has a time-independent and a fluctuating part (Chap.2). Obviously single particle excitations have energies independent of the wave vector and thus have the character of Einstein modes.

Still relatively little experimental work has been done on collective rotational excitations in molecular crystals. This is so for two reasons. 1) The measurement of dispersion curves requires single crystals which frequently are not available. 2) Because of several possible damping mechanisms the librational modes are often not well-defined excitations. A long review was published in 1970 by VENKATARAMAN and SAHNI /1.4/ and more recently a short one by DOLLING /1.5/.

In the present work the author will review the large field of single-particle rotations in molecular crystals. Some simple concepts will be introduced in Chap.2. In particular, the relation between rotational potentials and the type of rotational excitations found in molecular crystals will be discussed. We shall distinguish between classical diffusive rotations at high temperatures and quantum-mechanical rotations at low temperatures. Diffusion-like motion results if the fluctuating part of the potential is large. This is the case at high temperatures. At low temperatures the static part of the potential dominates and the rotational states can be calculated by solving a stationary Schrödinger equation.

Single-molecule rotations can be studied in a large class of molecular systems. Our definition of a molecular crystal will include both van der Waals and ionic crystals. A few digressions to 1) molecules diluted in matrices and 2) molecules adsorbed on surfaces will also be made.

In the same way as collective phenomena can be probed by coherent neutron scattering, single-particle rotations can be observed directly by incoherent neutron scattering. In coherent neutron scattering, the neutron waves originating from different scattering centres interfere and thus the phase relations between different molecules become important. This is not the case for "incoherent" neutron scattering. If the scattering length of a given kind of atoms varies statistically from position to position in the crystal there are no interference effects. The scattering then may be described as of independent scattering centres. Incoherent scattering results if there are various isotopes with different scattering lengths or if the scattering length depends on the orientation of the neutron spin in relation to that of the scattering nucleus (parallel and antiparallel). In our case the spin-dependent scattering is responsible for the observation of single-particle excitations. Fortunately, the most interesting molecules contain hydrogen atoms and protons happen to possess the largest known (spin) incoherent scattering cross section $\sigma_{inc}(H) = 79$ barn. In Chap.3 a derivation of the neutron scattering functions pertaining to the single-particle motion, both in the high-temperature and the low-temperature limit, will be given. In this chapter only general concepts will be introduced. More specific examples shall be discussed in Chaps.4 and 5 in connection with models describing the rotational motion.

Inelastic neutron scattering with monochromatization of the incoming neutrons (wave vector $\underline{k}$) and energy analysis of the scattered neutrons (wave vector $\underline{k}'$) yields information on the evolution of states in space and time. Measurements of the scattered neutron intensity can be performed both as a function of the momentum transfer $\underline{Q} = \underline{k}'-\underline{k}$ (typically $Q \cong 0.1$-8 Å^{-1}) and as a function of the energy transfer $E = (\hbar^2/2m_N)(\underline{k}^2-\underline{k}'^2)$. The former yields spatial information and the latter the excitation spectrum; m_N denotes the neutron mass. High-energy resolution has been a necessary condition for the success of the inelastic neutron scattering experiments described in the following. Only moderate $\underline{Q}$ resolution is required. The backscattering technique nowadays allows energy resolutions in the range between $\Delta E = 0.3$ µeV ($\cong$ 120 MHz) and 2 µeV, with energy transfers up to about 20 µeV. The range of energy transfers can be extended to several hundred µeV with time-of-flight and three-axis spectrometers. With use of long-wavelength neutrons, resolutions $\Delta E \cong 10$ µeV can be achieved. Concerning the design and the detailed characteristics of these spectrometers, the reader is referred to the literature (e.g., /1.6,7/ and the references therein).

When considering other techniques which have been successfully applied to the study of single-particle rotations, we must distinguish between high-temperature classical motion and rotations at low temperatures. Measurements of the spin-lattice relaxation time T_1 appear to be well-suited for the investigation of diffusive motions at high temperatures. A similar statement refers to optical measurements, and more specifically, to measurements of IR or Raman linewidths. Neither method yields information on the spatial aspects of molecular rotations, for example, on the geometry of rotational jumps. At low temperatures both specific-heat measurements and "advanced" NMR techniques have been used. Specific-heat measurements can supply information on the low-lying rotational states. A difficulty, however, lies in the connection between the rotational wave function and the nuclear-spin states of homonuclear molecules (for a detailed discussion we refer the reader to Chap.5). Practically all interesting transitions also involve changes of the nuclear-spin functions. This so-called spin conversion (a well-known example is the transition from ortho to parahydrogen) occurs on a time scale of about 1 to 10^5 seconds. In NMR, spectroscopic techniques recently have been developed. These are based on the concept of level crossing; nuclear-spin relaxation is speeded up when tuning a Zeemann splitting to resonance with rotational transitions. This is done by variation of the strength of an external magnetic field. If the Zeemann splitting of electrons is used, rotational transitions with energies up to several hundred eV can be observed. The electronic impurities which, in general, are created by irradiation of the sample (e.g., γ-irradiation of CH_4 produces free radicals with unpaired electrons) perturb the system and can yield spurious peaks. An alternative method consists of using the Zeemann splitting of the nuclear spins of the rotating molecules themselves. In this case energies $E \leqq 0.5$ µeV can be measured and thus information complementary to that from spin-dependent neutron scattering can be obtained.

The advantages of neutron scattering are its applicability to both high- and low-temperature rotation and the spatial information which none of the other techniques provides. First, classical diffusive rotation at high temperatures is discussed (Chap.4). This topic has already been treated in reviews by SPRINGER /1.8,9/ and LEADBETTER and LECHNER /1.10/. In particular /1.8/ represents an excellent introduction to translational and rotational diffusion and their investigation by quasi-elastic neutron scattering as well as by other techniques. The coverage of single-particle rotations seems to be incomplete without the inclusion of diffusive motions. In view of the above reviews, however, only a few very recent examples will be given.

So far only one relatively short review of single-particle rotations at low temperatures has been published /1.11/ and there has been considerable progress since. Therefore this aspect will be covered in more detail in Chaps.5-7. Chapter 5 is devoted to the calculation of rotational energies and wave functions of a molecule in a static rotational potential. For the calculation of transition matrix elements between two rotational states the inclusion of properly symmetrized nuclear-spin functions is essential. In Chap.6 a number of examples for the observation of low-temperature rotational states by spin-incoherent neutron scattering is given. An attempt is made to relate the observed spectra with the symmetry and magnitude of the respective single-particle potential. A distinction between almost free rotation and rotational tunneling is made. In the first case the rotational potential is weak and can be treated as a perturbation of the states of a free molecule. Rotational tunneling, on the other hand, is observed if the rotational potential is large, yet still allows a finite overlap of rotational wave functions in neighboring potential wells. The final chapter covers several special features of the low-temperature rotational motion. Merely by increasing the temperature a continuous transition to the regime of classical rotational motion is observed. The understanding of this temperature dependence seems to be of crucial importance for the understanding of single-particle rotations in general. The pressure dependence of tunneling states might allow learning about intermolecular interactions. Further topics include the isotope effect and the influence of substitutional impurities on tunneling states.

2. Interaction and Rotational Potentials

Before starting the discussion of molecular rotations in solids and their observation with neutron scattering, it appears useful to introduce a few simple aspects of molecular crystals and to give some clues for possible classifications. The discussion emphasizes the concept of single-particle potentials, their decomposition into a fluctuating and a static part, and the expansion into symmetry-adapted functions.

2.1 Rotational Degrees of Freedom

It is useful to list a catalogue of different situations which may be encountered in connection with molecular rotations. An almost trivial distinction is between one, two and three rotational degrees of freedom; examples are shown in Fig.2.1. The main difference concerns the complexity of an appropriate description which needs to keep track of just one angle in the one-dimensional case. Two angles are needed for linear molecules, while three angles (e.g., Euler angles or quaternions) are required for molecules of general shape. The constraint to rotation around just one axis may be due to two particularly large moments of inertia, covalent bonding (CH_3 group) or a dipole moment of the molecule (NH_3). Obviously, model calculations are simpler for one-dimensional rotors than, e.g., for tetrahedral molecules. On the other hand, the crystallographic structures of crystals composed of the latter are mostly better known and usually simpler.

2.2 Chemical Bonding

Molecular crystals may be further classified according to the nature of their chemical bonds. 1) Van der Waals-type of crystals consisting of neutral molecules (e.g., N_2) are the classical molecular crystals. 2) A second important class consists of

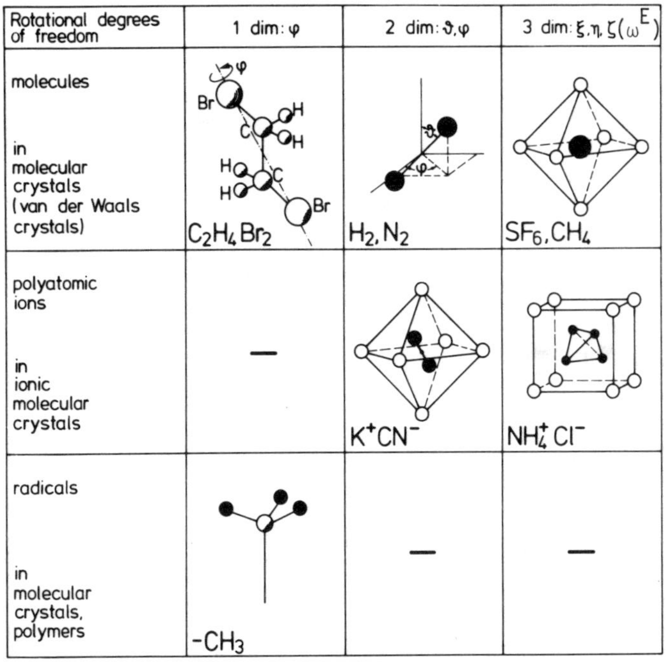

Fig.2.1. Examples for one, two, and three rotational degrees of freedom; $\omega^E = \xi, \eta, \zeta$ denotes the Euler angles which carry the axes fixed in a molecule from a standard orientation to any desired orientation by successive rotations of η about the z axis, ξ about the resulting y axis, and finally ζ about the resulting z axis

polyatomic ions in ionic crystals. 3) The importance of covalent bonding in defining one axis of rotation (σ bond: CH_3 groups) has already been mentioned above (Fig.2.1).

We shall mainly deal with small molecules or molecular groups. The chemical bonding within the molecules is then very strong, usually of covalent nature. This causes internal degrees of freedom—vibrational excitations, which deform the molecule—to have energies much higher than the external degrees of freedom, which comprise both translational and rotational motions. If these internal degrees of freedom may be neglected, the molecules may be considered as rigid units. This will be done in the following. Then the molecule can be described by a centre-of-mass coordinate $\underline{R}$ and its orientation, collectively denoted by ω^E.

2.3 Intermolecular Interaction

In contrast to dilute gases where the molecules rotate freely between collisions, they are subject to an angle-dependent potential in solids. If decomposed in terms of two-particle interactions between rigid molecules, two contributions may be distinguished /1.11/.

$$V_1^{ij} = V_1^{ij}(\omega_i^E;\underline{R}_i,\underline{R}_j) \qquad (2.1)$$

depends only on the orientation of the ith molecule as well as on the centre-of-mass (c.o.m.) positions $\underline{R}_j$;

$$V_2^{ij} = V_2^{ij}(\omega_i^E,\omega_j^E;\underline{R}_i,\underline{R}_j) \qquad (2.2)$$

on the other hand, depends on the orientations ω_i^E and ω_j^E of both interacting groups.

V_1^{ij} is the only contribution if a given molecule only interacts with particles without angular degrees of freedom, that is, atoms and monatomic ions. The term will be dominant for molecules in an atomic crystal matrix or in systems like NH_4X (X = Cl, Br, I) where the NH_4^+ ions are surrounded by halide ions and the octopole moment of the NH_4^+ groups interacts with the monopole moment of the halide ions /2.1-3/. Such a monopole-multipole part also exists in crystals consisting of neutral molecules. There the monopole is due to the angle-dependent van der Waals attraction and the hard-core repulsion averaged over the orientation of one of the two interacting partners /2.4/.

V_2^{ij} depends on the orientation of two interacting molecules and is responsible for all collective properties connected with orientation and rotation: orientational order-disorder phase transitions and orientational order as well as propagating librational excitations. V_2^{ij} can be phrased in terms of multipole-multipole interactions /2.5/. Examples are the quadrupole-quadrupole interaction in solid ortho-hydrogen /2.6/ or nitrogen and the octopole-octopole interaction in solid methane /2.7/. The interaction can be due to electrostatic interactions, hard-core repulsion, and van der Waals attraction. In this sense the notion of a multipole moment can be generalized beyond a strictly electrostatic meaning. Another, usually less important contribution, comes from the interaction of molecules via their polarizable neighbors.

2.4 Single Particle Potential

In the following we are mainly interested in the orientational potential acting on the ith molecule and obtained by summing V_1^{ij} and V_2^{ij} over all the neighbors of the ith molecule /1.11/.

$$W^i(\omega_i^E;\{\omega_j^{E'},\underline{R}_j\}) = \sum_{j=1}^{N+M}{}' V_1^{ij}(\omega_i^E;\underline{R}_i,\underline{R}_j) + \sum_{j=1}^{M}{}' V_2^{ij}(\omega_i^E,\omega_j^E;\underline{R}_i,\underline{R}_j) \ . \tag{2.3}$$

M and N denote the number of molecules and monatomic units in the crystal, respectively. The prime means that the term i = j should be omitted.

Other aspects, e.g., rotation-translation coupling /2.8/, become apparent after regrouping the terms in a different way. Here, we shall only include the definition of the crystalline field $V_c(\omega_i^E)$:

$$V_c(\omega_i^E) = \sum_j{}' V_1^{ij}(\omega_i^E;\underline{R}_i,\underline{R}_j)\Big|_{\underline{R}_i^0,\underline{R}_j^0} \ . \tag{2.4}$$

Here $\underline{R}_i^0$ and $\underline{R}_j^0$ denote equilibrium centre-of-mass positions.

W^i still depends on $\{\omega_j^{E'},\underline{R}_j\}$, which represents the set of centre-of-mass coordinates and angular coordinates of all other particles in the crystal. Obviously $\{\omega_j^{E'}(t),\underline{R}_j(r)\}$ is a time-dependent set of coordinates: within a classical picture $\underline{R}_j = \underline{R}_j(t)$ due to the lattice vibration in the crystal and $\omega_j^{E'} = \omega_j^{E'}(t)$ due to rotational excitations. Consequently W^i is time-dependent as well. Quantum mechanically $W^i(\omega_i^E,\{\omega_j^{E'};R_j\})$ needs to be integrated over the states of all the neighbors of the ith molecule and

$$V^i(\omega_i^E,t) = \int \prod_{j=1}^{N+M} d\underline{R}_j \prod_{j=1}^{N}{}' d\omega_j^E \ P(\{\omega_j^E,\underline{R}_j\},\{\omega_j^E,\underline{R}_j\};t) \ W^i(\omega_i^E,\{\omega_j^E,\underline{R}_j\}) \ . \tag{2.5}$$

Here $P(\{\omega_j^E;\underline{R}_j\},\{\omega_j^E;\underline{R}_j\}';t)$ is the density matrix of the molecular crystal. Again the prime in (2.5) means that integration over ω_j^E for j = 1 should be omitted. Usually it is impossible to handle the above expression and approximations are required, e.g., the replacement of P by a product of single particle density matrices (Hartree approximation /2.9/).

2.5 Classification of Single Particle Rotation

$V^i(\omega_i^E,t)$ may be decomposed into a static and a fluctuating part

$$V^i(\omega_i^E,t) = V_{st}^i(\omega_i^E) + V_{fl}^i(\omega_i^E,t) \quad . \tag{2.6}$$

V_{st}^i represents the time-dependent part of the potential, while the time average of $V_{fl}^i(\omega^E,t)$ vanishes for each ω_i^E. The fluctuations may be visualized as stochastic torques exerted on a molecule by its neighboring atoms and molecules. Their magnitude typically is of the order of kT.

Only the magnitude of the potential in relation to the rotational constant $B = \hbar^2/2\Theta$ (Θ = moment of inertia) is important and therefore a reduced dimensionless potential is introduced /2.10/

$$V'(\omega_i^E,t) = V_{st}(\omega_i^E)/B + V_{fl}(\omega_i^E,t)/B \ . \tag{2.7}$$

Depending on the relative magnitude of both the static and the fluctuating part of the scaled potential we may distinguish four characteristic situations which are listed in Table 2.1

Table 2.1. Classification of single particle rotations in molecular crystals in terms of the rotational potential. Different characteristic situations may be distinguished depending on the magnitude of the static and the fluctuating part of the potential

Static potential V_{st} \ Fluctuating potential V_{fl}	Strong	Weak
Strong	Rotational jumps	Librations and rotational tunneling
Weak	Rotational diffusion	Quantum-mechanical free rotation

High temperature means frequent transitions between the rotational states and phonon states in the crystal and thus a strong fluctuating part of the potential. In a number of orientationally disordered crystals of the van der Waals-type ("plastic crystals" like β-N_2 or CH_4I) the static or time-averaged part of the potential is rather weak. Then a diffusion of the molecules with respect to their angular degrees of freedom is taking place. This rotational diffusion is continuous in the angular variables, if V_{st} may be ignored completely. Otherwise both symmetry and magnitude of the potential have to be taken into account. In the limit of very strong static potentials V_{st}, the molecules are confined to a discrete number of equilibrium orientations which are occupied at random. Transitions across the barriers separating the equilibrium orientations occur by thermally activated jumps. In this limit the classical diffusive motion is called jump diffusion or molecular reorientation.

At low temperatures only few lattice phonon modes are populated and the system is close to its rotational ground state. Therefore we may expect the fluctuating part of the potential to be weak and a stationary quantum-mechanical picture should describe the situation rather well. Again two extremes may be distinguished. Because at low temperatures most systems are orientationally ordered, that with a strong ordering potential prevails. In this case one expects librational excitations of the molecules in their rotational potential, which in general is rather anharmonic. Additionally there is a tunnel splitting of these states. The splitting is due to the overlap of wave functions in neighboring potential wells. The other extreme, namely $B \geqq V^{i}(\omega^{E},t)$, i.e., $V' < 1$, is only rarely found in crystalline solids. The prime representative is solid hydrogen /2.6/, which is a quantum crystal, particularly with respect to its rotational degrees of freedom but also with respect to its translational degrees of freedom. Solid hydrogen always has been treated separately from all other molecular crystals, and an excellent review has recently been published /2.11/. Therefore there is no need for a detailed account of it here. There are, however, also systems with weak rotational potential at some sublattice sites with high symmetry which is due to a cancellation of the interactions with their neighbors. Examples are CH_4II /2.7,12/ and γ-O_2 /2.13/. Certainly we may find all sorts of intermediate situations between the aforementioned extremes. Of particular interest is the continuous transition between tunneling and rotational jumps which comes about merely by increasing temperature.

2.6 Static Orientational Potential V_{st}

2.6.1 Electrostatic Origin of V_{st}

In general the details and the relative magnitude of the various contributions to the rotational potential in a crystal are not well known. The situation is still relatively simple for electrostatic contributions in which case the knowledge of the leading multipole moments and the equilibrium distance suffices.

As an example we may take the electrostatic interaction between (for example) a tetrahedral molecule and its surroundings

$$V(\omega^E) = \iint \frac{\rho_T(\underline{r}')\rho_C(\underline{r})}{|\underline{r}'-\underline{r}|} d\underline{r}'d\underline{r} \; . \tag{2.8}$$

Here ω^E denotes the orientation of the molecule, $\rho_T(\underline{r}')$ its charge distribution, and $\rho_C(\underline{r})$ that of the crystal. One now may perform an expansion of the charge distributions into symmetry-adapted surface harmonics $K_{\ell m}(\theta,\phi)$ /2.14,15/ and

$$\rho_C(\underline{r}) = \sum_{\ell=0}^{\infty} \sum_{\mu=-\ell}^{\ell} b_{\ell\mu}(r) K_{\ell\mu}(\theta,\phi) \tag{2.9}$$

$$\rho_T(\underline{r}') = \sum_{\ell'=0}^{\infty} \sum_{\mu'=-\ell'}^{\ell'} a_{\ell'\mu'}(r') K_{\ell'\mu'}(\theta,\phi) \tag{2.10}$$

$$= \sum_{\ell'\mu'\mu''} a_{\ell'\mu'}(r') K_{\ell'\mu''}(\theta,\phi) U^{(\ell)}_{\mu''\mu'}(\omega^E) \tag{2.11}$$

with the expansion coefficients $b_{\ell\mu}(r)$ and $a_{\ell'\mu'}(r')$.

$\rho_C(\underline{r})$ is expanded in a (unprimed) frame fixed within the crystal, $\rho_T(\underline{r}')$ in a molecular (primed) frame, and the rotator functions /2.7,16,17/ $U^{(\ell)}_{mm'}(\omega^E)$ transform the surface harmonics from one frame to the other. For a tetrahedral molecule the lowest order harmonic $K_{\ell m}(\Omega)$ contributing to (2.9) is (apart from the angle-independent harmonic with $\ell = 0$)

$$K_{31}(\Omega) = \frac{\sqrt{105}}{4\pi} xyz \; . \tag{2.12}$$

The polar angles $\Omega = \theta,\phi$ are expressed in terms of $x = \sin\theta \cos\phi$, $y = \sin\theta \sin\phi$ and $z = \cos\theta$ which are coordinates on the surface of a three-dimensional unit sphere (in Sect.2.6.3 we replace x,y,z by τ_1,τ_2,τ_3). In methane, therefore, the leading interaction between two molecules (which is not what is presently considered; for the single particle potential we sum over the contributions from all surrounding molecules) is an octopole-octopole interaction /2.7/.

If, additionally, $1/|\underline{r}-\underline{r}'|$ is expanded for $r' < r$

$$\frac{1}{|\underline{r}-\underline{r}'|} = 4\pi \sum_{\ell''m} \frac{1}{2\ell''+1} \frac{r'^{\ell''}}{r^{\ell''+1}} K_{\ell''m}(\Omega_r) K^*_{\ell''m}(\Omega_{r'}) \tag{2.13}$$

and if use of the orthogonality of the functions $K_{\ell m}$ is made when performing the integrations in (2.8), one obtains

$$V(\omega^E) = 4\pi \iint dr'dr \sum_{\ell\mu\mu'} a_{\ell\mu'}(r')\, b_{\ell\mu}(r) \frac{r'^{\ell}}{(2\ell+1)r^{\ell+1}} U^{(\ell)}_{\mu\mu'}(\omega^E)$$

$$= \sum_{\ell\mu\mu'} B^{(\ell)}_{\mu\mu'}\, U^{(\ell)}_{\mu\mu'}(\omega^E) \; . \tag{2.14}$$

Obviously both the molecular *and* the site symmetry, see (2.9,10), determine which of the coefficients $B^{(\ell)}_{\mu\mu'}$ are nonzero. An expression for a tetrahedral molecule in a potential of tetrahedral symmetry will be given later (Fig.2.2).

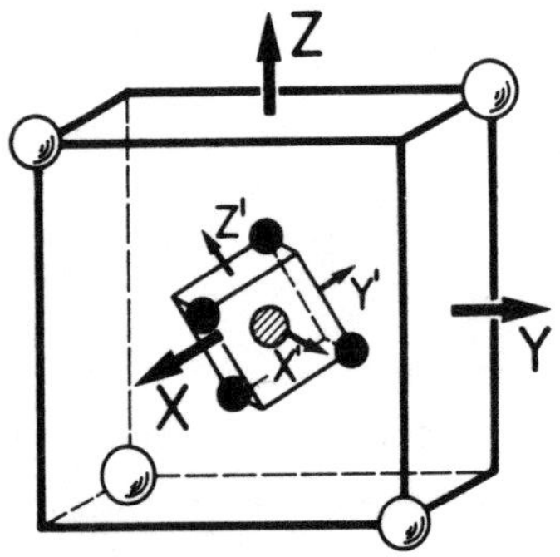

Fig.2.2. Tetrahedron in a potential of tetrahedral symmetry

2.6.2 Valence and Dispersion Forces

For anisotropic dispersion forces as well as for valence forces it is much more difficult to give explicit expressions /2.4,5/. Usually phenomenological potentials are taken, mostly atom-atom potentials $v(r)$ /2.18/ or, e.g., the Kihara core potential /2.19,20/. The phenomenological parameters between pairs of atoms often are chosen to be independent of the chemical bonding of the individual atom /2.20/. This approach has had some success, particularly in the description of organic crystals with only very few parameters. It is, in general, inadequate if microscopic behavior such as phonon dispersion or rotational tunneling is analysed in these terms.

The formulation given above for electrostatic interactions with $v \sim 1/r$ /2.21/ has been generalized /2.4/ to power laws $v \sim 1/r^n$. The result has been used to calculate both the octopole-octopole interaction and the crystalline field in methane starting from 6-12 potentials /2.5/. The radial dependence of the single particle potential now contains terms $V(\omega^E) \sim 1/r^{\ell+n}$ with $n = 6,12$ and $\ell = 4,6$ ...

244966

BLACKWELL'S SHOP/PHONE ORDER
(Delete as appropriate)

BHB ORDER No. 8/JACT/114

ept: SCI | Initials: RI | Date: 10.11.84

ame and Address:

I. JACKSON
St. Johns C.
Ox.

Notes:
(including name of person placing order, if order is for an Institute)

end To (if different from above)

15/12

ADVISE ✓	NON ACCOUNT CUSTOMER		ACCOUNT CUSTOMER FIRM ORDER
	Part Payment £	Till Receipt No.	Uncollected books will be forwarded

SEND AND CHARGE	Account Number	CASH SALE INVOICE ☐ Tick here
	CREDIT Type: CARD	Tel. No: Number:

Number of Copies	Author, Title, Series etc	Part Pymt.	Style	Publisher
1	PRESS, W.	*	PB	~~Springer~~
	"Molecular Rotations in Solids"			12-11-84.
	* no deposit taken			

ACTION →	Reserved	To Inv.	To CWO	To Cr. Card	To Stock Dist	Advised	Answered	Refunded
DATE →					3/12			

98

2.6.3 Expansion into Symmetry-Adapted Functions with One, Two and Three Angular Degrees of Freedom

In spite of successes in the prediction of the orientational order in CD_4II /2.7,22/ or the calculation of rotational excitations in ammonium salts /2.16,23/, the knowledge of intermolecular interactions still is not satisfactory. Therefore very often the single particle potential simply is expanded into a set of orthonormal functions. The expansion coefficients then are taken as adjustable parameters, which may be compared with parameters obtained from model pair potentials. In order to learn details of the intermolecular interactions from such a comparison it is necessary to perform measurements as a function of pressure (e.g., of the tunnel splitting); the most direct access to $V[\omega^E(r)]$ is via its dependence on the equilibrium distance, which changes with pressure (Sect.7.2).

In the following we give the expansion of the potential into a complete set of symmetry-adapted surface harmonics for one, two and three angular degrees of freedom (the latter has already been done in Sect.2.6.1).

1) If the molecular rotation is confined to just one angular degree of freedom ϕ, the potential can be expanded into a series of trigonometric functions

$$V_{st}(\phi) = \sum_{n=1}^{\infty} (a_n \cos n\phi + b_n \sin n\phi) \ . \tag{2.15}$$

The role of symmetry can be illustrated by the example of the one-dimensional rotation of CH_3 or NH_3 groups around the threefold symmetry axis of the molecules (Fig.2.3). Rotations around this axis must leave the potential unchanged. Therefore the molecular symmetry only allows nonzero terms with $n = 3,6,9,\ldots$ An additional reduction results, if the site symmetry is not a subgroup of the molecular symmetry /2.24/. A simple example is provided by the presence of a mirror plane (which contains the rotation axis). It causes $b_n = 0$ for all n and the corresponding symmetry-adapted functions are $\cos 3n\phi$. If the site symmetry provides a twofold or fourfold axis in addition to the threefold axis of the molecule, orientational disorder results. In the high-temperature phase of $Ni(NH_3)_6I_2$ the crystal symmetry along the axis of rotation introduces both a mirror plane and a fourfold rotation axis. As a consequence the molecules are orientationally disordered and all coefficients except a_{12}, a_{24} ... vanish /2.25/.

In order to obtain a unified representation for all three dimensions it is useful to introduce the angular coordinates as coordinates on a unit circle, $\tau_1 = \cos\phi$, and $\tau_2 = \sin\phi$, and expand V_{st} into these coordinates

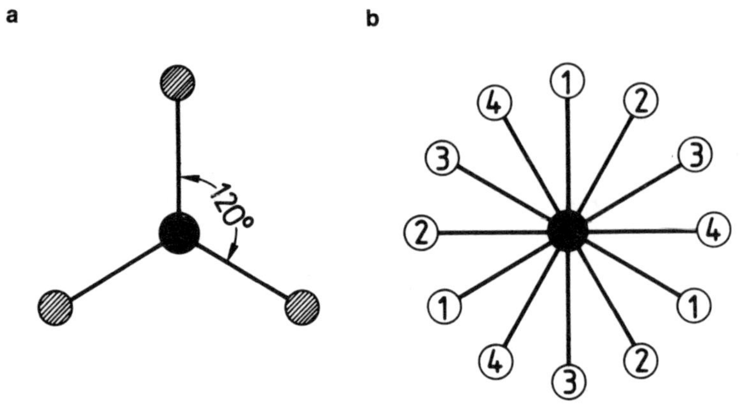

Fig.2.3. (a) View of a CH_3 group or a NH_3 molecule along the threefold symmetry axis of the molecule (schematically). (b) Four equivalent orientations of these molecules (viewed as above) originating from the presence of a fourfold axis at the lattice site. Each of the four orientations has the same statistical weight

$$V_{st}(\phi) = \sum_{n=1}^{\infty} (c_n\tau_1^n + d_n\tau_2^n) \; . \tag{2.16}$$

2) Linear molecules possess two angular degrees of freedom $\Omega = (\theta,\phi)$. In this case an expansion of the potential into a series of spherical harmonics is appropriate

$$V_{st}(\theta,\phi) = \sum_{\ell=0}^{\infty} \sum_{m=-\ell}^{\ell} a_{\ell m} Y_{\ell m}(\theta,\phi) \; . \tag{2.17}$$

The transition to symmetry-adapted harmonics is very simple if either the molecule or its site processes a centre of inversion. Due to the relation $Y_{\ell m}(\Omega) = (-1)^{\ell} Y_{\ell m}(-\Omega)$ all terms with odd ℓ vanish (e.g., for β-N_2 /2.26,27/. Also, if there is just one axis of symmetry, this imposes simple conditions on the allowed indices m. In a case of high symmetry it is especially advantageous (particularly for the 5 cubic point groups) to use symmetry-adapted harmonics. For cubic symmetry, examples being the alkali cyanides in their orientationally disordered fcc phase /2.28/, the expansion reads

$$V_{st}(\theta,\phi) = \sum_{\ell=0}^{\infty} \sum_{m=1}^{2\ell+1} b_{\ell m} K_{\ell m}(\theta,\phi) \; . \tag{2.18}$$

Only a few coefficients $b_{\ell m}$ are nonzero. In case of cubic symmetry (point group m3m) these are b_{01}, b_{41}, b_{61}, b_{81} ... The cubic harmonics $K_{\ell m}(\theta,\phi)$ usually are expressed in terms of coordinates on the three-dimensional unit sphere τ_1, τ_2, and τ_3, which have been introduced before. An explicit expression for $\ell = 4$ is [the lowest order term for tetrahedral symmetry is given in (2.9)]

$$K_{41}(\tau_1,\tau_2,\tau_3) = \frac{1}{4\pi}\frac{5}{4}\sqrt{21}\,(\tau_1^4+\tau_2^4+\tau_3^4-0.6) \tag{2.19}$$

A potential V_{st} of cubic symmetry truncated after $\ell = 4$ often is referred to as the "Devonshire potential" /2.29/. It has six minima (or potential pockets) along [100] for b_{41} negative and eight minima along [111] for b_{41} positive.

3) Two- and three-dimensional molecules like C_6H_6 or CH_4 have three angular degrees of freedom. A standard choice for three angular coordinated are the Euler angles $\omega^E = (\xi,\eta,\zeta)$. They carry the axis fixed in a molecule from a standard orientation to any desired orientation by successive rotations of η about the z axis, ξ about the resulting y axis and, finally, ζ about the resulting z axis. If no particular symmetry is specified, an appropriate set of orthonormal functions is provided by the Wigner $\mathcal{D}$ functions and

$$V_{st}(\xi,\eta,\zeta) = \sum_{\ell=0}^{\infty}\ \sum_{m,m'=-\ell}^{\ell} A^{(\ell)}_{mm'}\,\mathcal{D}^{(\ell)}_{mm'}(\xi,\eta,\zeta)\ . \tag{2.20}$$

As the number of parameters increases with $(2\ell+1)^2$ the introduction of symmetry-adapted rotator functions is especially helpful (2.14).

For tetrahedral molecules at a site of tetrahedral symmetry, many coefficients $B^{(\ell)}_{\mu\mu'}$ vanish and up to order $\ell = 6$ only $B^{(0)}_{11}$, $B^{(3)}_{11}$, $B^{(4)}_{11}$ and $B^{(6)}_{11}$ are nonzero. Even more than in the previous cases it is useful to introduce coordinates on the four-dimensional unit sphere, the quaternions $\tau = \tau_1,\tau_2,\tau_3,\tau_4$ with $\tau_1^2+\tau_2^2+\tau_3^2+\tau_4^2 = 1$ /2.16,30/. τ and $-\tau$ denote the same set of angles. The quaternions have a simple geometrical meaning; $\tau_1 = x_R\sin(\phi/2)$, $\tau_2 = y_R\sin(\phi/2)$, $\tau_3 = z_R\sin(\phi/2)$, $\tau_4 = \cos(\phi/2)$, where ϕ denotes the angle of rotations around the axis $\underline{x}_R = (x_R, y_R, z_R)$ in a Cartesian coordinate system. In general their use is much simpler than that of the Euler angles ω^E. The relation to the Euler angles is

$$\begin{aligned}
\tau_1 &= \sin\tfrac{1}{2}\eta\,\sin\tfrac{1}{2}(\xi-\zeta)\\
\tau_2 &= \sin\tfrac{1}{2}\eta\,\cos\tfrac{1}{2}(\xi-\zeta)\\
\tau_3 &= \cos\tfrac{1}{2}\eta\,\sin\tfrac{1}{2}(\xi+\zeta)\\
\tau_4 &= \cos\tfrac{1}{2}\eta\,\cos\tfrac{1}{2}(\xi+\zeta)
\end{aligned} \tag{2.21}$$

and the $U^{(\ell)}_{\mu\mu'}(\omega)$ are converted into rotator functions $H^{(\ell)}_{\mu\mu'}(\tau)$ which are polynomials in the τ_i of order 2ℓ /2.7,16,17/ and

$$V(\tau) = B^{(3)}_{11}H^{(3)}_{11}(\tau) + B^{(4)}_{11}H^{(4)}_{11} + B^{(6)}_{11}H^{(6)}_{11} + \dots \tag{2.22}$$

As an example we give the leading terms for point symmetry $\overline{4}3m$ (Fig.2.2)

$$H^{(3)}_{11}(\tau) = 16(\tau_1^6 + \tau_2^6 + \tau_3^6 + \tau_4^6) - 20(\tau_1^4 + \tau_2^4 + \tau_3^4 + \tau_4^4) + 5 \ . \tag{2.23}$$

$H^{(3)}_{11}(\tau)$ is invariant against 192 symmetry operations (e.g., 4! from permutations of τ_i). As with the surface harmonics there are no odd-order terms in the presence of a centre of symmetry and therefore $H^{(4)}_{11}(\omega^E)$ is the first term for point symmetry m3m.

3. Neutron Scattering

In this chapter the neutron scattering from rotating molecules shall be discussed. For this purpose it is necessary to introduce some aspects of the theory of neutron scattering /3.1/. This will be done without derivation of fundamental expressions such as (3.1,16). Here the reader is referred to standard textbooks as, e.g., /3.1/. Naturally the emphasis is on the (spin-dependent) incoherent neutron scattering which is related to single particle properties. Single particle properties are observed if the spin-dependent scattering lengths of atoms of a given kind but at different positions in the crystal are statistically independent. An equivalent formulation in terms of transition matrix elements (Chap.5) can be given for molecules. The validity of this statistical independence must be examined in specific cases.

There are two alternative formulations of the neutron scattering law. One is based on transition matrix elements between quantum-mechanical states. It will be used in connection with low-temperature rotational motion. The other formulation uses classical self-correlation functions within the concept developed by van HOVE /3.2/. It applies for the diffusive rotational motion at high temperatures. In this chapter only the principal ideas shall be developed. It appears useful to give specific examples only after the underlying model for the classical rotational motions (Chap.4) or rotational states (Chap.5) have been formulated.

3.1 Formulation Based on Transition Matrix Elements

First we will introduce the partial differential cross section $d^2\sigma/d\Omega d\varepsilon$ ($d\Omega$ = solid angle element, $\hbar d\varepsilon$ = energy interval) which results from scattering of neutrons treated in the first Born approximation. The transition probability of the total system from an initial state $|\psi_a \mu \underline{k}\rangle$ to a final state $|\psi_{a'} \mu' \underline{k}'\rangle$ is calculated, and the first Born approximation means that this is done in first-order perturbation theory. $|\psi_a\rangle$ denotes the wave function of the scatterer, $|\mu \underline{k}\rangle = |\mu\rangle \exp(i\underline{k}\underline{r})$ the spin state and wave vector of the neutron plane wave. Unprimed and primed symbols refer to quantities before and after the scattering, respectively.

We will follow rather closely the notation in /3.3/. The double differential cross section then is

$$\frac{d^2\sigma}{d\Omega d\varepsilon} = \frac{k'}{k} \sum_{\mu\mu'} \sum_{aa'} p_\mu p_a |\langle\psi_{a'}\mu' k'|W|\psi_a \mu \underline{k}\rangle|^2 \, \delta(\omega-\omega_{aa'}) \; . \qquad (3.1)$$

p_μ is the statistical weight for the initial state of the neutron. For unpolarized neutrons the number of spin-up neutrons $|\mu\rangle = |\alpha\rangle$ and spin-down neutrons $|\beta\rangle$ is the same and $p_\alpha = p_\beta = 1/2$. p_a denotes the population of the initial state of the scattering molecule. $\hbar\omega_{aa'}$ is the energy difference $E_a - E_{a'}$ between the initial and the final state of the scatterer. Finally W denotes the spin-dependent nuclear interaction between the neutron and the scatterer, expressed in terms of the Fermi pseudo-potential.

$$W = \sum_{n=1}^{N} \sum_{\gamma=1}^{N} A^{n\gamma} \, \delta(\underline{r}-\underline{r}_{n\gamma}) \qquad (3.2)$$

$$A^{n\gamma} = a^{\gamma}_{coh} + [2a^{\gamma}_{inc}/\sqrt{I(I+1)}] \; \mathring{s}\cdot\mathring{i}_{n\gamma} \; . \qquad (3.3)$$

Here $\underline{r}_{n\gamma}$ denotes the position of the γth atom (proton!) in the nth molecule, $\mathring{i}_n$ its spin operator and I its total spin. N is the number of molecules in the crystal and there are M atoms in the molecule. Correspondingly, $\underline{r}$ and $\mathring{s}$ denote the position and the spin-operator, respectively, of the neutron. a_{inc} is the spin-dependent part of the scattering length a, a_{coh} is spin-independent. The connection with the scattering length a_+ and a_- for the spins of neutron and nucleus parallel and antiparallel, respectively, is (with the index γ dropped)

$$a_{coh} = \frac{1}{2I+1} [(I+1)a_+ + Ia_-] \qquad (3.4)$$

$$a_{inc} = \frac{\sqrt{I(I+1)}}{2I+1} (a_+ - a_-) \; . \qquad (3.5)$$

As the spin incoherence is not a property of the scattering nucleus, but a consequence of the statistical independence of the spin states in the crystal, the notations a_{coh} and a_{inc} which are found in textbooks should be replaced by a_{sd} and a_{si}; here sd stands for spin-dependent and si for spin-independent.

In the following it is assumed that the single-molecule states of different molecules are uncorrelated (Chap.5). Then the spin states of scattering atoms belonging to two different molecules ($n_1 \neq n_2$) are not correlated, either. The double differential cross section /3.2/ is

$$\frac{d^2\sigma}{d\Omega d\varepsilon} = \frac{k'}{k} \sum_{\mu\mu'} \sum_{aa'} p_\mu p_a \sum_{\substack{n_1\gamma_1 \\ n_2\gamma_2}} \langle\mu\psi_a|A^{n_1\gamma_1}\exp(-i\underline{Q}\cdot\underline{r}_{n_1\gamma_1})|\mu'\psi_{a'}\rangle \cdot \langle\mu'\psi_a|A^{n_2\gamma_2}\exp(i\underline{Q}\cdot\underline{r}_{n_2\gamma_2})|\mu\psi_a\rangle\,\delta(\omega-\omega_{aa'}) \;. \tag{3.6}$$

$\underline{Q} = \underline{k}'-\underline{k}$ is the momentum transfer of the neutrons. For $n_1 \neq n_2$ (absence of correlations) the matrix elements of $A^{n_1\gamma_1}$ and $A^{n_2\gamma_2}$ may be replaced by their averages $\bar{A}$, which allows the following separation:

$$\frac{d^2\sigma}{d\Omega d\varepsilon} = \frac{d^2\sigma_d}{d\Omega d\varepsilon} + \frac{d^2\sigma_s}{d\Omega d\varepsilon} \;. \tag{3.7}$$

$d^2\sigma_d/d\Omega d\varepsilon$ does not contain any spin dependence (d denotes "distinct")

$$\frac{d^2\sigma_d}{d\Omega d\varepsilon} = \frac{k'}{k} \sum_{\mu\mu'} \sum_{aa'} p_\mu p_a \bar{A}^2 \sum_{\substack{n_1\gamma_1 n_2\gamma_2 \\ n_1 \neq n_2}} \langle\mu\psi_a|\exp(-i\underline{Q}\cdot\underline{r}_{n_1\gamma_1})|\mu'\psi_{a'}\rangle \cdot \langle\mu'\psi_{a'}|\exp(i\underline{Q}\cdot\underline{r}_{n_2\gamma_2})|\mu\psi_a\rangle\,\delta(\omega-\omega_{aa'}) \;. \tag{3.8}$$

It represents the coherent scattering and as such is connected with the collective properties of the system (e.g., the periodic structure in a crystal leads to Bragg peaks).

$\bar{A}$ is obtained by averaging over all the spin states $|\nu\rangle$ of the (homonuclear) atoms in a molecule and

$$\bar{A} = \sum_{\mu\mu'} \sum_{\nu\nu'} p_\mu p_\nu \frac{1}{NM} \sum_{n\gamma} \langle\nu'\mu'|A^{n\gamma}|\nu\rangle \;. \tag{3.9}$$

For an unpolarized target $\bar{A} = a_{coh.}$ On the other hand

$$\frac{d^2\sigma_s}{d\Omega d\varepsilon} = \sum_{\mu\mu'}\sum_{aa'} p_\mu p_a \sum_{\substack{n\\ \gamma_1\gamma_2}} \langle\mu\psi_a|(A^{n\gamma_1}-\bar{A})\exp(-i\underset{\sim}{Q}\cdot\underset{\sim}{r}_{n\gamma_1})|\mu'\psi_{a'}\rangle$$

$$\cdot \langle\mu'\psi_{a'}|(A^{n\gamma_2}-\bar{A})\exp(i\underset{\sim}{Q}\cdot\underset{\sim}{r}_{n\gamma_2})|\mu\psi_a\rangle\,\delta(\omega-\omega_{aa'}) \tag{3.10}$$

represents the incoherent scattering, which bears information on the single molecule aspects (s denotes "self").

$$A^{n\gamma}-\bar{A} = \left[2a_{inc}/\sqrt{I(I+1)}\right]\mathring{s}\cdot\mathring{i}_{n\gamma} \ . \tag{3.11}$$

In order to proceed further it is necessary to calculate molecular wave functions (which comprise both a rotational and a spin part, see Chap.5) and apply

$$\mathring{s}\cdot\mathring{i}_{n\gamma} = \mathring{s}_z\mathring{i}_{n\gamma z} + \frac{1}{2}\left(\mathring{s}_+\mathring{i}_{n\gamma-} + \mathring{s}_-\mathring{i}_{n\gamma+}\right) \tag{3.12}$$

in order to calculate the matrix element /3.3,4/. $\mathring{s}_+$, $\mathring{i}_{n\gamma+}$ and $\mathring{s}_-$, $\mathring{i}_{n\gamma-}$ denote creation and annihilation operators, respectively; $\mathring{s}_z$ and $\mathring{i}_z$ the z components of the operators $\mathring{s}$ and $\mathring{i}$. The action of the second term in (3.12), (1/2) $\mathring{s}_+\mathring{i}_{n\gamma-}$, e.g., is as follows: $i_{n\gamma-}$ reduces the z component of the nth nuclear spin by $\Delta I_z = 1$ while $\mathring{s}_+$ increases the z component of the neutron spin by $\Delta I_z = 1$. The nuclear spins are correlated within a molecule (see Chap.5), giving rise to interference effects at low temperatures. Therefore the scattering is not strictly incoherent (in which case the nuclear spins have to be uncorrelated). However, as long as the spin-states of two different molecules are uncorrelated, the interference is restricted to within a molecule and one still observes single-molecule motions.

As mentioned in the introduction the best candidates for successful experiments are protonated molecules, because the spin-dependence of the scattering from protons is particularly strong. Very recently low-energy rotational excitations were also observed for deuterated molecules (Sect.7.3).

3.2 Formulation Based on Classical Self-Correlation Functions

An alternative formulation in terms of time-dependent atomic coordinates has been given by van HOVE /3.2/. Again the scattering is separated into a coherent and an incoherent part

$$\frac{d^2\sigma}{d\Omega d\varepsilon} = \frac{k'}{k}\left[4\pi a_{coh}^2 S_{coh}(\underset{\sim}{Q},\omega) + 4\pi a_{inc}^2 S_{inc}(\underset{\sim}{Q},\omega)\right] . \qquad (3.13)$$

In contrast to the definitions in the previous section the scattering length here has been extracted from the scattering function. This is strictly correct only for a monatomic crystal, but represents a very good approximation if the scattering from one atomic species dominates. In general this is true for samples containing protons. For reasons of brevity we will only give expressions connecting with the incoherent (single particle) scattering. Usually it is assumed that the spins of different nuclei (also within a molecule) are uncorrelated at high temperatures. The validity of this assumption, that correlations within a molecule can be ignored, has been investigated in /3.2,5,6/.

$S(\underset{\sim}{Q},\omega)$ is the fundamental quantity which is determined in a neutron scattering experiment. It is i) real, ii) fulfills detailed balance

$$S(\underset{\sim}{Q},\omega) = \exp(\hbar\omega/kT)\, S(-\underset{\sim}{Q},-\omega) \; , \qquad (3.14)$$

and iii) obeys sum rules for $\int d\omega\, \omega^n\, S(\underset{\sim}{Q},\omega)$ /3.1,2/. As indicated, we only want to deal with the incoherent scattering function $S_{inc}(\underset{\sim}{Q},\omega)$ which can be introduced as the Fourier transform of an intermediate scattering function $I_S(\underset{\sim}{Q},t)$

$$S_{inc}(\underset{\sim}{Q},\omega) = \frac{1}{2\pi}\int_{-\infty}^{\infty} \exp(-i\omega t)\, I_S(\underset{\sim}{Q},t)dt \qquad (3.15)$$

and

$$I_S(\underset{\sim}{Q},t) = \frac{1}{N}\sum_{i=1}^{N} \langle \exp[-i\underset{\sim}{Q}\cdot\underset{\sim}{r}_i(0)]\, \exp[i\underset{\sim}{Q}\cdot\underset{\sim}{r}_i(t)]\rangle \; . \qquad (3.16)$$

The brackets denote a thermal average; N is the number of nuclei; the quantities $\underset{\sim}{r}_i(t)$ and $\underset{\sim}{r}_i(0)$ are introduced as quantum-mechanical operators which do not commute. In this case the self-correlation function $G_S(\underset{\sim}{r},\underset{\sim}{t})$, which has been introduced by van Hove, is a complex function and has no simple physical meaning

$$S_{inc}(\underset{\sim}{Q},\omega) = \frac{1}{2\pi}\int \exp[i(\underset{\sim}{Q}\cdot\underset{\sim}{r}-\omega t)]\; G_S(\underset{\sim}{r},t)d\underset{\sim}{r}dt \; . \qquad (3.17)$$

The van HOVE formalism /3.2/ usually is applied in the classical high-temperature regime, where the operators commute. Then $G_S(\underset{\sim}{r},t)$ is the probability of finding a given particle at the position $\underset{\sim}{r}$ at time t, if it was at $\underset{\sim}{r} = 0$ at time $t = 0$. We will stick to a classical meaning of $G_S(\underset{\sim}{r},t)$ throughout. It is advantageous, however, to introduce a generalized self-correlation function /1.10/ $G_S(\underset{\sim}{r},\underset{\sim}{r}_0;t)$ which is the probability of finding a given particle at $\underline{r}$ at time t, if it was at $\underline{r}_0$ at t = 0 [obviously $G_S(\underline{r};0;t) \equiv G_S(\underline{r},t)$]. The probability distribution of the initial positions $\underline{r}_0$ is given by $g(\underline{r}_0)$ and then the thermal average can be written as

$$I_S(\underline{Q},t) = \iint \exp[i\underline{Q}(\underline{r}-\underline{r}_0)]\, G_S(\underline{r},\underline{r}_0;t)\, g(\underline{r}_0) d\underline{r} d\underline{r}_0 \quad . \tag{3.18}$$

In order to obtain $S_{inc}(\underline{Q},\omega)$ for a polycrystal, an average over the angular coordinates Ω_Q of $\underline{Q}$ has to be performed

$$S_{inc}(Q,\omega)|_p = \frac{1}{4\pi} \int d\Omega_Q S_{inc}(\underline{Q},\omega) \quad . \tag{3.19}$$

A simultaneous treatment of all kinds of motion a molecule can perform usually is impossible. Therefore approximations are required. In general it is assumed that the various kinds of motion are uncorrelated, that is, that internal vibrations of the molecule and translational and rotational motions happen independently. In this case the intermediate scattering function separates into a product.

$$I_S(\underline{Q},t) = \prod_i I_S^{(i)}(\underline{Q},t) \tag{3.20}$$

where i enumerates the various degrees of freedom. For example, I_S^T stands for the translational and I_S^R for the rotational function. As internal vibrational energies of molecules are usually of the order of 100 meV and above, they are much larger than the energies connected with the external degrees of freedom ($\lesssim$ 10 meV). Therefore the molecules can be considered as rigid and the internal vibrations are ignored. Furthermore, the centre-of-mass motion (coordinate $\underline{R}$) is considered independent of the rotational motion (coordinate $\underline{\rho} = \underline{r}-\underline{R}$). This approximation is not always justified; the importance of a coupling between translational and rotational degrees of freedom in some systems has recently been shown /2.8/. It is particularly important in case of strongly anisotropic molecules, and obviously the approximation is much better justified for globular molecules. So far a formulation of this coupling has only been given for simple probability distribution functions /3.7/ which are directly connected with $G_S(\underline{r},t\rightarrow\infty)$, as will be shown later. No attempt to include such a coupling in the description of classical diffusive motions of a molecule in a molecular crystal has so far been made.

The translational function $I_S^T(\underline{Q},t)$ can be written in terms of a phonon expansion /1.10/ of the intermediate scattering functions

$$I_S^T(Q,t) = \exp(-2W)\, [1+I_1(\underline{Q},t) + \ldots] \quad . \tag{3.21}$$

The first term represents the elastic scattering, the second term the one-phonon scattering, the third term the two-phonon scattering, and so on. The exp(-2W) is the Debye-Waller factor and for isotropic mean-squared amplitudes $\langle u^2\rangle$ the Debye-Waller factor simply is $\exp(-\langle u^2\rangle Q^2)$. Now it is assumed that the scattering can be separated into a low-energy quasielastic part due to the diffusive motion and a high-energy part due to the lattice vibrations. The quasielastic scattering function then reads

$$I_S(\underline{Q},t) = I_S^R(\underline{Q},t)\exp(-2W) \quad . \qquad (3.22)$$

The acoustic phonons cause some problems, in particular at large momentum transfers. Recently LOTTNER et al. /3.8/ have shown that the acoustic modes, if approximated by a Debye spectrum, contribute a flat background to the quasielastic scattering which increases with temperatures. The effect of the diffusive motion on the phonon scattering function has been neglected in this consideration.

The main problem remains, namely, the evaluation of the rotational intermediate scattering function

$$I_S^R(\underline{Q},t) = \iint \exp[i\underline{Q}(\underline{r}(t)-\underline{r}_0)]\, G_S(\underline{r},\underline{r}_0;t)\, g(\underline{r}_0)d\underline{r}_0 d\underline{r} \quad . \qquad (3.23)$$

As $\underline{r}$ is restricted in space (e.g., to the surface of a sphere) some authors have introduced the rotational analogue of the van Hove self-correlation function $G(\omega_0^E,\omega^E;t)$ /1.10,3.9,10/. It is the conditional probability of finding a molecule with orientation ω^E at time t, given that the orientation was ω_0^E at time t = 0

$$I_S^R(\underline{Q},t) = \iint \exp[i\underline{Q}(\underline{r}(t)-\underline{r}_0)]\, G(\omega^E,\omega_0^E;t)\, f(\omega_0^E)d\omega_0^E d\omega^E \quad . \qquad (3.24)$$

Here $f(\omega_0^E)$ represents the probability distribution function of initial orientations ω_0^E. Specific cases in which certain model assumptions allow the calculation of $G(\omega^E,\omega_0^E;t)$ are discussed in the next section.

Some additional remarks pertaining to the elastic scattering, i.e., to $S(\underline{Q},\omega=0)$ can be made independent of the particular model chosen for the diffusion mechanism. Afterwards we return to $S(\underline{Q},\omega)$. For infinitely long time t /1.10,3.2/

$$\lim G_S^R(\underline{r},\underline{r}_0;t) = g^R(\underline{r}_0) \qquad (3.25)$$

$$I_S^R(\underline{Q},t=\infty) = \left|\int \exp(i\underline{Q}\cdot\underline{r})\, g^R(\underline{r})d\underline{r}\right|^2 \qquad (3.26)$$

and

$$S_{inc}(\underline{Q},\omega)|_{el} = I_S^R(\underline{Q},t=\infty)\,\delta(\omega) \quad . \tag{3.27}$$

$I_S^R(\underline{Q},\infty)$ sometimes is called the elastic incoherent structure factor (EISF). A δ function is present in (3.27) because $G_S^R(\underline{r},\underline{r}_0;t)$ does not decay to zero for $t\to\infty$. This is different from the case of translational diffusion. There an atom or a molecule can spread out over the whole crystal and $g_R^R(\underline{r}_0)\to 0$. In contrast to this, $\underline{r}$ is restricted in space for rotations and $g^R(\underline{r})$ remains finite. Being directly connected with the Fourier transform of the probability distribution function $g^R(\underline{r})$, structural information is contained in the EISF. Its knowledge can be very useful for the distinction between models with different jump geometry. In principle the EISF might yield information superior to that provided by Bragg scattering as it may be observed continuously in reciprocal space as a function of $\underline{Q}$. In practice it is very difficult 1) to separate the EISF from other sources of elastic scattering, 2) to separate it from the quasielastic scattering, and 3) to correct for multiple scattering /3.11,12/. As a rule, measurements have to be extended to $Q\rho > \pi/2$ ($\rho = |\underline{\rho}|$ = distance of atoms from molecular c.o.m.), if a distinction between various models should be made.

Classical self-correlation functions $G_S(\underline{r},t)$ lead to symmetric scattering functions $S_S(\underline{Q},\omega) = S_S(\underline{Q},-\omega)$ which obviously do not fulfill the condition of detailed balance (3.14). Therefore it is customary to include a detailed balance factor and to define (see, e.g., /1.8,10/.

$$S_{inc}(\underline{Q},\omega) = \exp(\hbar\omega/2kT)\,S_S(\underline{Q},\omega) \quad . \tag{3.28}$$

The scattering function so obtained obeys the detailed balance condition (3.14). Its first moment (sum rules!) diverges, however. As long as the energy width of the quasielastic scattering $\Gamma \ll kT$, this poses no serious problems. Difficulties may arise, however, in diluted molecular systems, which remain orientationally disordered down to very low temperatures. In this case formulations in analogy to scattering functions of paramagnetic salts /3.1/ may provide a solution.

4. Stochastic Rotational Motion

The diffusion of atoms in gases or liquids is a well-known phenomenon. In the solid state sizable translational diffusion is encountered at temperatures close to the melting point and over a wider temperature range in ionic conductors. In an analoguous way there is also diffusion-like motion of molecules with respect to the angular degrees of freedom. As is the case with translational diffusion, the rotational motion of interacting particles at high temperatures in principle represents an N-body problem. Practicable descriptions of the diffusion process obviously require approximations. These and resulting models for the high-temperature rotational motion are discussed in the following.

4.1 General Aspects

In an approximate picture one describes the motion of a given particle in the force field of its neighbors /4.1/. This force field will usually consist of a static and a fluctuating part. The static part $\underline{f}(\omega^E)$ is the negative gradient of the static rotational potential (here ω^E collectively denotes angular coordinates again). The fluctuating force field $\underline{F}(t)$ can be decomposed into a part $\underline{f}(\omega^E,t)$ the ensemble average of which vanishes and a friction part of the form $-\Theta\zeta_0\dot{\omega}^E(t)$. It is proportional to the angular velocity $\dot{\omega}^E(t)$; ζ_0 is a friction constant. If it is assumed that the fluctuating force $\underline{f}(t)$ is uncorrelated with both the angular velocity and with $\underline{f}(t=0)$ one arrives at the Langevin theory of Brownian motion and the following equation (the formulation given below is symbolic, if ω^E denotes more than one angular degree of freedom):

$$\Theta\ddot{\omega}^E(t) = -\Theta\zeta_0\dot{\omega}^E(t) + \underline{f}_0(\omega^E) + \underline{f}(\omega^E;t) \ . \qquad (4.1)$$

The above approximations are correct for large particles immersed in a system of small particles. Frequent collisions cause fluctuations on a much shorter time scale than typical time constants of the particle under consideration. Certainly, this is not true in a molecular crystal in which all particles are of comparable size.

One can go beyond the Langevin equation by i) introducing memory effects into the friction term or by ii) adjusting the short time behavior to the real situation in a crystal (for references see /1.8/). Often it is assumed /4.2/ that for short times τ_1 the particles perform an oscillatory motion, then diffuse for a time τ_2, etc. The calculations, however, ignore the static force $\underline{f}_0(\omega;t)$ arising, for example, from the crystal field. In most cases the neglect of $\underline{f}_0(\omega^E;t)$ probably represents an even more serious approximation than the restriction to a Brownian type of motion. Only very recently, with the availability of high-energy resolution and the use of multiple scattering corrections, the quality of neutron data has improved to a point where the limitations of the Langevin theory (for rotational motions) could be checked. So far even the distinction between fairly simple models has turned out difficult. Concerning models beyond the Langevin theory the reader is therefore referred to /1.8/ which gives an excellent review of the whole topic. All these models rather apply to molecular liquids than to molecular crystals.

In practice, models have been utilized which are based on Langevin diffusion plus further approximations. Mostly two extremes have been used in the data analysis: 1) The static potential and consequently the static force $\underline{f}_0(\omega^E)$ in (4.1) have been neglected. This leads to rotational Brownian motion in the absence of a potential. There are only few molecular crystals for which this assumption can safely be applied. 2) The opposite limit is to take very large potentials giving rise to a strong orientational localization of the molecules in their potential wells. The molecules change their orientation by thermal activation across the potential barrier. The path of a molecule in configuration space is ignored and the rate of jumps to and from a given site is chosen to describe the stochastic angular motion. In the following, we will restrict ourselves to the limiting cases, particularly concerning the experimental examples. In the last part of this chapter, however, we shall return to models which describe the diffusion in a potential and which seem to describe the rotational diffusion in molecular crystals better.

4.2 Continuous Rotational Diffusion

As mentioned before, most stochastic models deal with the diffusion in the absence of a periodic potential. This leads to significant simplifications since the probability distribution functions $g^R(\underline{r}_0)$ or, equivalently, $f(\omega_0^E)$ are angle independent. On the other hand one must admit that there are only very few molecular solids where a complete neglect of the static potential V_{st} seems to be permissible. Therefore it appears that the models within the limit $V_{st} = 0$ are somewhat better suited for the description of molecular liquids. There are, however, a few orientationally disordered crystals for which rotational diffusion provides a reasonable approximation. Examples, like the rotation of the n-paraffins /4.3/, will be given later.

We will restrict ourselves to problems in which just one value of the moment of inertia Θ enters. Consequently there is just one rotational constant B and only one diffusion constant D_R. There are three such cases (see also Sect.2.1).

1) Uniaxial rotation of CH_3 groups or NH_3 molecules. In both cases the rotation around two of the three axes is quenched. 2) Linear molecules have only two rotational degrees of freedom and the two rotational constants are equal. 3) Obviously the values of the three moments of inertia are identical for spherical top molecules. For all three cases the self-correlation function $G(\omega^E,\omega_0^E;t)$ only depends on the "difference" of the orientations $\omega^E-\omega_0^E$ [more strictly one should write $\omega^E(\omega_0^E)^{-1}$; defined by rotational operators $\hat{R}$ acting on a reference orientation ω_1^E one has $\omega^E = \hat{R}\omega_1^E$, $\omega_0^E = \hat{R}_0\omega_1^E$ and consequently $(\omega^E-\omega_0^E) = \hat{R}\hat{R}_0^{-1}\omega_1^E = \omega^E(\omega_0^E)^{-1}$]. $G(\omega^E,\omega_0^E;t)$ can be expanded into a double sum of rotator functions of the appropriate dimension. For a spherical top, e.g., one obtains /3.10/

$$G(\omega^E,\omega_0^E;t) = \sum_{\ell mm'} \frac{2\ell+1}{8\pi^2} F^{\ell}_{m'm}(t) \sum_{\mu} \mathcal{D}^{(\ell)}_{m'\mu}(\omega^E)\, \mathcal{D}^{(\ell)}_{m\mu}(\omega_0^E) \quad ; \tag{4.2}$$

with the initial condition $G(\omega^E,\omega_0^E;t=0) = \delta[\omega^E(\omega_0^E)^{-1}]$ it is found that only relaxation functions $F^{\ell}_{00}(t) \equiv F_{\ell}(t)$ (with $m=m'=0$) contribute. $\int G(\omega^E,\omega_0^E;t)\,d\omega^E = 1$ necessitates $F_0(t) = 1$ for all times t and the initial condition requires $F_{\ell}(0) = 1$ for all orders ℓ /4.2/.

Langevin diffusion is characterized by two limits. Here it is useful to introduce $\tau_R = 1/\zeta_0$ as a characteristic time. For short times $\tau \ll \tau_R$, one has a free particle behavior and $\langle r^2(t)\rangle = \int r^2 G_s(\underline{r},t)dr \sim t^2$. For $t > \tau_R$, on the other hand, one obtains the diffusion limit and $\langle r^2(t)\rangle = D_R(t-\tau_R)/6$; here τ_R represents something like a delay time, after which diffusion starts. The simplest and one of the most frequently used assumptions is that the molecule diffuses at all times. This means the neglect of the delay time τ_R which is permissible only if the friction constant ζ_0 is large. Then the Langevin equation leads to a self-correlation function which obeys a diffusion equation

$$D_R\Delta_{\omega}G(\omega^E,\omega_0^E;t) = \partial G(\omega^E,\omega_0^E;t)/\partial t \tag{4.3}$$

with the following solution for the relaxation functions F_{ℓ}, introduced above (4.2):

$$F_\ell(t) = \exp[-\ell(\ell+1)D_R t] \ . \tag{4.4}$$

D_R is the rotational diffusion constant and $D_R = \langle\alpha^2\rangle/6\tau$ relates it with the magnitude α of the diffusive angular steps. The intermediate scattering function is

$$I_S^R(\underset{\sim}{Q},t) = \sum_{\ell=0}^{\infty} (2\ell+1) j_\ell^2(Q\rho) F_\ell(t) \tag{4.5}$$

and after Fourier transforming with respect to time

$$S_{inc}^R(\underset{\sim}{Q},\omega) = j_0(Q\rho)\delta(\omega) + \sum_{\ell=0}^{\infty} (2\ell+1) j_\ell(Q\rho) \frac{\Gamma_\ell/\pi}{\omega^2+\Gamma_\ell^2} \tag{4.6}$$

with $\Gamma_\ell = \ell(\ell+1)D_R$. The elastic incoherent structure factor (EISF) is directly related with $I_R^S(Q,t=\infty)$. For continuous rotational diffusion all $F_\ell(t)$ vanish for $t\to\infty$ (4.4), except $F_0(t)$. Therefore the EISF simply is given by the spherical Bessel function $j_0(Q\rho)$ which is the Fourier transform for a spherical shell. This is generally true for $V_{st} = 0$, as long as $F_\ell(t)$ is based on some kind of diffusive motion. The quasielastic scattering consists of an infinite sum of Lorentzians of width Γ_ℓ. This differs from reorientation models (see Sect.4.4) where only a discrete number of Lorentzians contributes and the width $\Gamma(\underset{\sim}{Q})$ remains finite. For continuous rotational diffusion $S(\underset{\sim}{Q},\omega)$ obviously remains unchanged after a powder average is performed as it only depends on $|\underset{\sim}{Q}|$.

The same calculation for uniaxial rotational diffusion /4.4/, e.g., for a CH_3 group, yields

$$S_{inc}^R(\underset{\sim}{Q},\omega) = J_0^2(Q\rho\sin\theta)\delta(\omega) + 2\sum_{\ell=1}^{\infty} J_\ell^2(Q\rho\sin\theta) \frac{\Gamma_\ell/\pi}{\Gamma_\ell^2+\omega^2} \tag{4.7}$$

where $\Gamma_\ell = D_R\cdot\ell^2$. In order to obtain a polycrystalline average the Bessel functions (of the first kind) $J_\ell^2(Q\rho\sin\theta)$ have to be replaced by

$$\langle J_\ell^2(Q\rho)\rangle = \frac{1}{2}\int_{-1}^{1} J_\ell(Q\rho\sin\theta)\, d\cos\theta;$$

here θ denotes the angle between $\underset{\sim}{Q}$ and the axis of rotation. Here and in the following the scattering is given per proton.

If τ_R is not neglected, that is, if the Langevin equation in absence of a static force is used, a somewhat modified relaxation function results (e.g., /1.8,4.5/).

$$F_\ell(t) = \exp\{\ell(\ell+1)D_R\tau_R[1-t/\tau_R-\exp(-t/\tau_R)]\} \ . \tag{4.8}$$

Another generalization can be obtained if a single diffusive step α is replaced by a distribution function $W(\alpha)$ /4.6/, e.g., a Gaussian distribution.

Results for classical rotation in absence of friction can be obtained from quantum-mechanical calculations by going to the classical limit ($kT \gg \hbar^2/2\Theta$). In this case the coupling between spatial wave functions and nuclear-spin functions—which is important at low temperatures—can be ignored and the calculation of neutron scattering transition matrix elements becomes much simpler (Sect.5.3). Such calculations have been performed for freely rotating molecules by SEARS /3.9,10,4.7/ and de RAEDT /4.8/. They provide very useful tests of stochastic models in the weak friction limit.

"More-advanced" models generally necessitate the introduction of additional parameters and one has to check carefully what generalization is most meaningful. One example in which discriminating between two different models proved impossible is reported in /4.9/. Both in neopentane and in t-butyl chloride Langevin rotational diffusion and a two-step stochastic model fitted the data equally well over a wide temperature range. This was attributed to very short relaxation times. It may be expected that new efforts, particularly with single crystalline samples, will be undertaken to achieve more insight.

4.3 Examples for the Limit of Rotational Diffusion

Two simple examples of uniaxial rotational diffusion are discussed in the following. In the high-temperature structures of $Ni(NH_3)_6I_2$ /2.25,4.10/ and of the paraffin $C_{33}H_{68}$ /4.11/ the limit in which the static potential V_{st} can be neglected seems to be approached rather closely.

4.3.1 $Ni(NH_3)_6I_2$

The arrangement of the NH_3 groups in the cubic antifluorite structure of $Ni(NH_3)_6I_2$ ($T > T_c \cong 20$ K) is shown in Fig.4.1. The Ni ions are surrounded by octahedrally coordinated NH_3 groups, the dipole moments of which are aligned along the cubic axes. The static periodic potential experienced by the molecules must be invariant against both the molecular (3m) and the site symmetry (4/mmm). The combination of the threefold axis of the molecule and the fourfold cubic axis leads to an effective potential of the form

$$V(\phi) = \sum_{\ell=1}^{\infty} V_{12\ell} \cos 12\ell\phi \qquad (4.9)$$

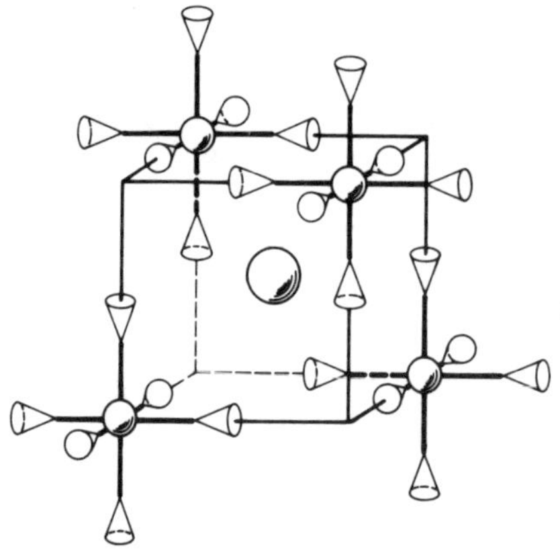

Fig.4.1. High temperature structure of $Ni(NH_3)_6I_2$ (space group Fm3m). Only the four $[Ni(NH_3)_6]^{++}$ complexes at the origin and at the face centers of the unit cell are shown. The orientationally disordered NH_3 groups are represented by cones

which is rapidly oscillating with ϕ. The resultant orientational disorder either can be sketched as in Fig.2.2b or, more realistically, as the modulation of an angle-independent density distribution by terms of the type $a_\ell \cos 12\ell\phi$. The correction to the elastic incoherent structure factor in (4.7) to the lowest order is $C_{12}J_{12}(Q\rho\sin\phi)$ or the corresponding averaged quantity. Such deviations become sizable only for momentum transfers $Q \gtrsim 8$ Å^{-1} and therefore for all practical purposes can be neglected. Similarly we may assume that such a potential has little effect on the dynamical properties.

Measurements have been performed with powder samples of $Ni(NH_3)_6I_2$ /2.25/. In order to test the applicability of different models it is particularly useful to look into the Q dependence of the quasielastic scattering. Figure 4.2 shows the elastic intensity as extracted from measurements at T = 90 K in the range 0.5 Å^{-1} $\leq Q \leq$ 4.5 Å^{-1}. The experimental results are compared to the EISF for two simple models: uniaxial rotational diffusion and 120° jumps (the latter model has been used to analyze quasielastic scattering in $Ni(NH_3)_6(ClO_4)_2$ /4.4/. Agreement is found with the diffusion model only. From Fig.4.2 it is evident that even extremely different models yield an almost identical EISF for $Q\rho \leq 2$ and measurements need to be extended beyond this limit. This does not come as a surprise. The EISF is proportional to $|F(\underline{Q})|^2$, where $F(\underline{Q})$ is the formfactor of the density distribution generated by a single diffusing proton (3.26). Formally the same formfactor $F(\underline{Q})$ is obtained when calculating the coherent Bragg scattering from a crystal composed by molecules. Then $F(\underline{Q})$ describes interferences of the neutron waves originating from different atoms (this distinguishes coherent from incoherent scattering). Due to this analogy the same methods can be used which have been developed for the analysis of orientationally disordered structures /4.12-14/ on the basis of coherent Bragg scattering. There the angle-dependent part of the density distribution at a

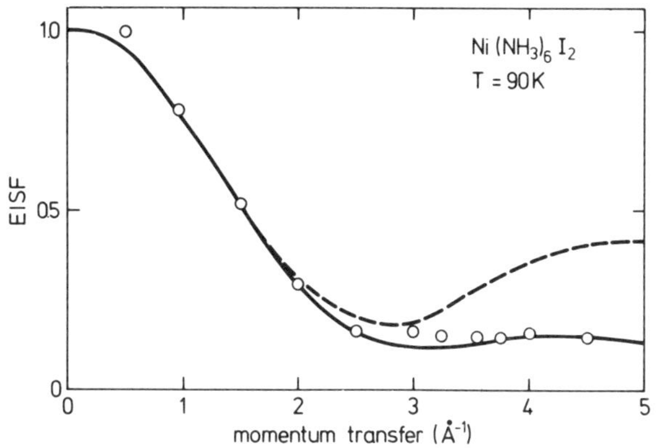

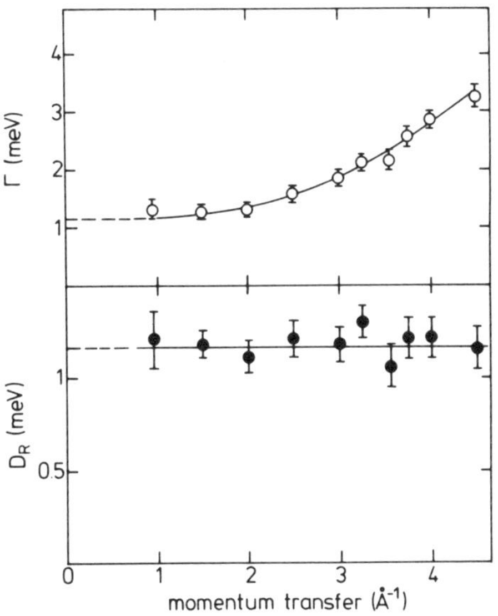

▲ Fig.4.2. Measurement of the elastic incoherent structure factor (EISF) in $Ni(NH_3)_6I_2$ at T = 90 K are compared with calculations based on one-dimensional rotational diffusion (——) and 120° jumps (----) /2.25/

Fig.4.3. Analysis of quasielastic neutron scattering spectra in $Ni(NH_3)_6I_2$ at 90 K with a single Lorentzian of width Γ (120° jump model) and the rotational diffusion model (lower half of the figure) /2.25/. The latter model yields a Q-independent rotational diffusion constant D_R and therefore seems to represent the diffusive motion much better than the jump model

molecular site is expanded into symmetry-adapted surface harmonics (for uniaxial rotors, trigonometric functions). In case of continuous diffusion there is no angle dependence and the EISF simply is $\langle J_0(Q\rho)^2\rangle$. For the 120° jump model there are also trigonometric functions with argument $3n\phi$. Therefore the structure factor additionally contains Bessel functions of order ℓ = 3, 6, 9, ... These contribute significantly only for large momentum transfer and a distinction between different models cannot be based on measurements at small momentum transfer Q.

The difference between the aforementioned two models becomes even more evident when the quasielastic part of the scattering is analyzed. Rotational jump models (see Sect.4.4) always predict only a finite number of Lorentzians, each with a Q independent width. For uniaxial rotation with a restriction to 120° jumps the situation is particularly simple, as there is just one Lorentzian. For continuous rotational diffusion, on the other hand, the calculation leads to an infinite series of Lorentzians with increasing width [see (4.7)]. One now analyzes the observed spectra with a single Lorentzian. If the linewidth is Q independent, the jump model holds, which is not true otherwise. Figure 4.3 shows that the width so obtained increases monotonously with Q and, therefore, the jump model must be discarded. The spectra measured at different Q values can, however, be explained with a single diffusion constant D_R, as may be seen in Fig.4.3.

It is interesting to look at the temperature dependence of the diffusion constant D_R. Figure 4.4 shows spectra of measurements at four different temperatures. The spectrum at T = 5 K is taken at a temperature well below the orientational order-disorder transition temperature T_0 = 19.9 K in $Ni(NH_3)_6I_2$. In this phase the molecules become orientationally ordered, experience a potential with a leading term $V_3\cos3\phi$ and display rotational tunneling. Tunneling lines could not be resolved with an energy resolution of about 0.30 meV [used for all scans aimed to determine $D_R(T)$]. Better resolution yields a tunnel splitting of about 65 μeV. It may be noted that at 5 K there is practically no inelastic scattering found in the range $0.1 \le |E| \le 2$ meV. An analysis of the quasielastic scattering (above the phase transition) in

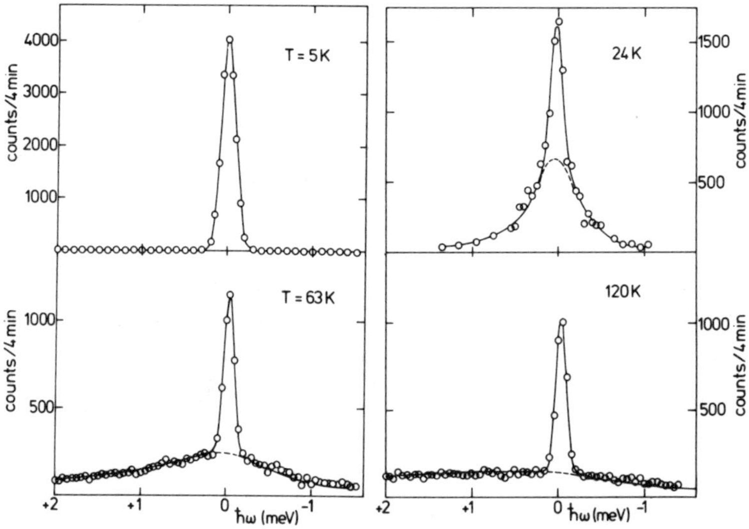

Fig.4.4. Temperature dependence of the quasielastic neutron scattering from $Ni(NH_3)_6I_2$ /2.25/. The measurement at T = 5 K is below the order-disorder phase transition (T_0 = 19.9 K) and can be used to determine the background level

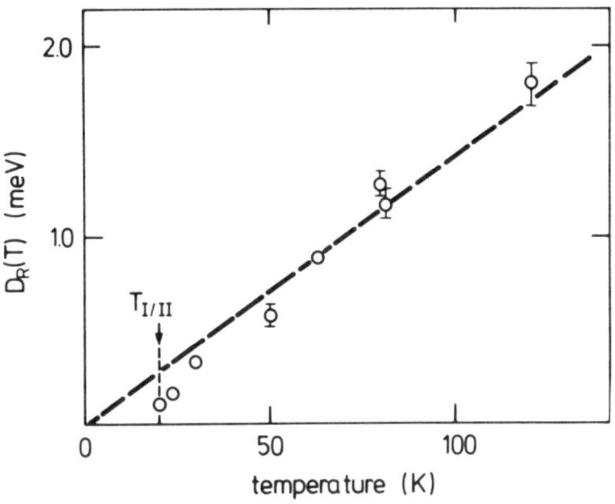

Fig.4.5. Rotational diffusion constant $D_R(T)$ as obtained from a fit of the continuous rotational diffusion model to the spectra from $Ni(NH_3)_6I_2$ at various temperatures /2.25/. The dashed line shows the Einstein behavior, $D_R(T) \sim kT$. Deviations from this line near the phase transition are probably due to orientational correlations between NH_3 groups

terms of continuous rotational diffusion leads to the results displayed in Fig.4.5. The diffusion constant D_R decreases almost linearly for $50 \leqq T \leqq 130$ K and obeys the Einstein relation $D_R = kT/(\zeta_0\Theta)$; ζ_0 = friction constant. Approaching the phase transition, D_R deviates increasingly from this relation to the side of lower values. This may have two reasons. 1) It may reflect the effect of a small but finite activation energy $E^A \cong 25$ K. 2) The deviations may signal the breakdown of single particle diffusion in the vicinity of the phase transition temperature T_0. The orientational correlations slow down critically on approaching T_0 (the transition is of first order, but starts like a continuous transition). This may show up in the single particle rotations as well, as they represent a weighted average of all rotational modes in the crystal. Detailed calculations are needed in order to relate the observed deviations near T_0 to orientational correlations or to discard this picture.

4.3.2 n-Paraffins ($C_{33}H_{68}$)

Another class of systems in which uniaxial rotational diffusion has been found are the n-paraffins. Stochastic rotational motions have been studied in paraffins of different length—mainly by the method of quasielastic neutron scattering /4.3,15,16/. Here only one recent work shall be reported, which was performed with a single crystal of $C_{33}H_{68}$ /4.3/. Four stable phases of $C_{33}H_{68}$ exist between room temperature and the melting point at $T_m = 71.8$ °C /4.11/. In all phases the straight chain paraffin molecules (Fig.4.6a)—lenght ∿45 Å—form lamellae-like structures as schematically shown in Fig.4.6b. The various phases differ with respect to the translational and rotational disorder of the molecules and consequently also different types of stochastic motions are characteristic for each phase /4.3,17/. We will only discuss phase D, which is stable in the narrow temperature range 68 °C ≦ T ≦ 71.8 °C. In

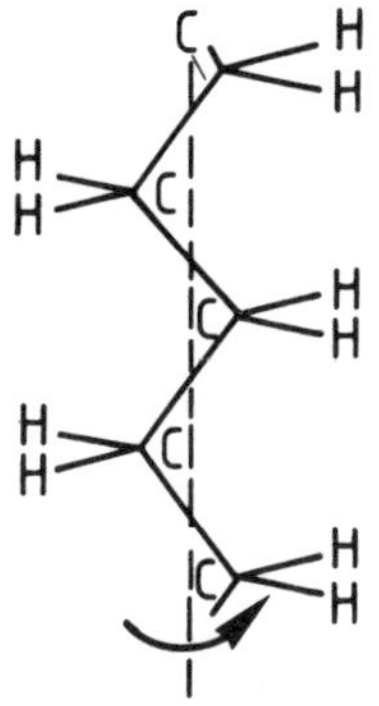

a)

Fig.4.6. (a) Section of straight chain paraffin (schematic); (b) schematic representation of the structural properties of the n-alkane $C_{33}H_{68}$ in the four solid phases A,B,C and D /4.11/

modification	A	B	C	D
schematic structure	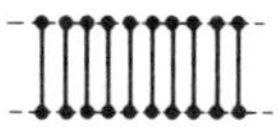	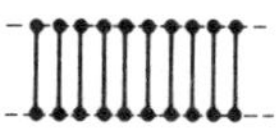	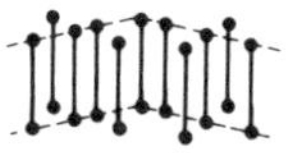	
crystal class	orthorhombic	monoclinic	monoclinic	monoclinic (triclinic)
state of translational order	no interfacial defects	no interfacial defects	irregular staggering of straight chains	intrachain defects
type of rotational motion	90° - jumps of single molecules	cooperative 180° - jumps	flip - flop - motion	rotation, kink diffusion

b)

this phase the measurement of temperature-dependent diffusion constants as has been performed for $Ni(NH_3)_6I_2$ is not meaningful. Phase D has a strongly disordered structure: i) orientationally it is probably close to complete disorder; ii) translational disorder shows up in the form of extended interfacial defects (several Å long) giving rise to rough surfaces of the lamellae; iii) additionally there are intrachain defects in the form of kinks which locally cause deviations from the straight chain configuration. Quasielastic scattering studies with energy resolution of 1, 8, and 40 μeV (using different spectrometers) have been performed with the momentum transfer parallel ($Q_{\parallel}$) and perpendicular ($Q_{\perp}$) to the chain direction. The latter allows a separation between translational diffusion and rotational diffusion, a problem that is also met in similar form in connection with liquid crystals.

There has been no evidence for kink diffusion at all. Concerning the angular motion the low symmetry in the crystal does not immediately suggest complete orientational disorder (in contrast to $Ni(NH_3)_6I_2$ /2.25/). Obviously, the rotational motion has to be uniaxial, but there is not necessarily uniaxial rotational diffusion. In case of $C_{33}H_{68}$, rotational diffusion of "rigid rods" may be concluded from the EISF and from the $Q_{\perp}$-dependent width of the quasielastic spectra. Kinks probably are present, but kink motion is not observed on the time-scale defined by the energy resolution of the experiment /4.3/. A rotational diffusion constant $D_R = 6.10^{10}$ s^{-1} has been found which is somewhat smaller than the values reported for shorter paraffin chains /4.15,16/. Translational diffusion over an extension of 4.5 ± 0.5 Å – which seems to define the length-scale of the interfacial defects – and a translational diffusion constant $D_T = 1.0 \cdot 10^5$ cm^2/s are found. All fits have included multiple scattering corrections /4.3/.

The number of crystals in which rotational diffusion may be expected to describe the stochastic rotational motion is relatively small. Examples with more than one rotational degree of freedom are β-N_2 (coherent scatterer; molecular dynamics calculations are available /4.18/) and CH_4 I or CH_4 diluted in rare-gas solid /4.19/.

4.4 Rotational Jump Model

In case of a strong static rotational potential one is in a limit opposite to the one of rotational diffusion. A molecule librates around any one of a finite number of allowed equilibrium orientations, before it changes to a different equilibrium orientation in a diffusive step (e.g., a 120° jump for a CH_3 group). The rotational jump model /4.20,21/ assumes that the jumps are instantaneous, that is, the jump time required for a diffusive step can be neglected in comparison with the time between consecutive jumps (residence time τ). It is not obvious that the assumption of instantaneous jumps is always justified (even for strong orientational localization). In particular, it seems dangerous to estimate jump times assuming that a molecule rotates freely (classically free) on its way from one minimum to the next. The molecule either has to pass through a saddle-point or a maximum of the rotational potential and as this will consume a large part of its kinetic energy, the flight time is much longer. In another approximation which is always used, the centre-of-mass motion is treated as statistically independent from the rotational motion.

Assuming instantaneous jumps the rotational self-correlation function reads /1.10/

$$G_S(\underline{r},\underline{r}_0;t) = \sum_{j=1}^{M} p_j^{(i)}(t)\ \delta[\underline{r}-(\underline{r}_{i0}-\underline{r}_j)] \ . \tag{4.10}$$

M is the number of allowed sites of a given proton, and $p_j^{(i)}(t)$ denotes the probability of finding a proton at the site $\underline{r}_j$ at time t, provided that it was at $\underline{r}_{i0}$ at time t = 0. With the formulation given in (4.10), the integration over all initial positions $\int d\underline{r}_0$ in (3.23) is replaced by a summation over a discrete number of positions $\underline{r}_{i0}$. The δ function in (4.10) certainly can be replaced by a Gaussian which may account for a finite librational amplitude.

If the successive jumps are uncorrelated the probabilities $p_j^{(i)}(t)$ obey a simple system of coupled differential equations, sometimes referred to as rate equations

$$dp_j^{(i)}(t)/dt = \frac{1}{\tau}\left[\frac{1}{M'}\sum_{\lambda=1}^{M'} p_\lambda^{(i)}(t) - p_j^{(i)}(t)\right] . \tag{4.11}$$

The sum includes all M' sites which are accessible via jumps from the site $\underline{r}_j$ and all jumps from such sites back to $\underline{r}_j$. For simplicity it has been assumed that the equilibrium occupation of the sites is constant and that all allowed decay channels are equally probable. The probability for a jump of a proton is $w = 1/\tau$, where τ is the average time between two successive jumps. Before solving (4.11) it is necessary to specify the initial conditions $p_j^{(i)}(t=0) = \delta_{ij}$ and the normalization

$$\sum_{j=1}^{M} p_j^{(i)} = 1 \text{ for all i.}$$

In the following some of the steps on the way to the scattering function $S^R_{inc}(\underline{Q},\omega)$ will be illustrated with the example of an equilaterial triangle rotating around its axis of symmetry (CH_3 group, NH_3, ...). In this case there are three sites $\underline{r}_1' = (1,0,0)\rho_0$, $\underline{r}_2' = (-1/2, \sqrt{3}/2, 0)\rho_0$, $r_3' = (-1/2, -\sqrt{3}/2, 0)\rho_0$ and only 120° jumps are possible. Here ρ_0 denotes the distance of the atoms from the threefold symmetry axis. The rate equations (index i omitted, $w = \tau^{-1}$) are as follows:

$$\begin{aligned} \dot{p}_1 &= -wp_1 + \tfrac{1}{2}wp_2 + \tfrac{1}{2}wp_3 \\ \dot{p}_2 &= \tfrac{1}{2}wp_1 - wp_2 + \tfrac{1}{2}wp_3 \\ \dot{p}_3 &= \tfrac{1}{2}wp_1 + \tfrac{1}{2}wp_2 - wp_3 . \end{aligned} \tag{4.12}$$

Sometimes a modified meaning for w is found in the literature; w' denotes the probability of jumps to a distinct site and — for the present geometry — $w' = w/2$.

With the ansatz $\underline{p} = \underline{q}\exp(\lambda t)$ one obtains

$$\lambda\underline{q} = \overline{\overline{W}}\underline{q} \quad \text{with} \quad \overline{\overline{W}} = \frac{w}{2}\begin{pmatrix} -2 & 1 & 1 \\ 1 & -2 & 1 \\ 1 & 1 & -2 \end{pmatrix} . \tag{4.13}$$

The symmetry of the problem is reflected in the eigenvectors of $\overline{\overline{W}}$. In the present case this is the group of even permutations of three particles which is isomorphous with the group C_3 (3-fold rotation axis). The eigenvalue $\lambda_1 = 0$ belongs to the completely symmetric (stationary) solution with $\underline{q}_1 = \frac{1}{3}(1,1,1)$ and immediately yields the elastic incoherent structure factor (EISF). The eigenvalues $\lambda_{2/3} = -\frac{3}{2}w$, on the other hand, belong to the solution with E symmetry. The eigenvector is $\underline{q} = \frac{1}{3}(1,\varepsilon,\varepsilon^*)$ and $\underline{q}_3$ is the complex conjugate of $\underline{q}_2$; $\varepsilon = \frac{1}{2} + i\sqrt{3}/2$ and ε^* are cube roots of 1. Obviously completely symmetric combinations of the $p_j^i(t)$ do not decay with time while combinations of lower symmetry do decay. While not really necessary in the present example, group theoretical arguments become very helpful in case of a large number of sites and also, if both molecular and site symmetry are important /4.22-26/.

With use of the initial conditions and the normalization one obtains

$$\begin{aligned} p(t) = p_1(t) &= \frac{1}{3} + \frac{2}{3}\exp(-\frac{3}{2}wt) \\ p_2(t) = p_3(t) &= \frac{1}{2}[1-p(t)] \ . \end{aligned} \tag{4.14}$$

Taking $\underline{r}_1 = 0$ as origin and with $\underline{r}_{12} = \underline{r}_2' - \underline{r}_1'$, $\underline{r}_{13} = \underline{r}_3' - \underline{r}_1'$ the self-correlation function reads

$$G_S^R(\underline{r},\underline{r}_1;t) = \delta(\underline{r})p(t) + [\delta(\underline{r}-\underline{r}_{12}) + \delta(\underline{r}-\underline{r}_{13})]\,[1-p(t)]/2 \ . \tag{4.15}$$

Corresponding expressions are obtained, if $\underline{r}_2$ and $\underline{r}_3$, respectively, are taken as origin and then

$$I_S^R(\underline{Q},t) = p(t) + \frac{1}{3}[1-p(t)]\,A(\underline{Q}) \tag{4.16}$$

with

$$A(\underline{Q}) = \cos\underline{Q}\cdot\underline{r}_{12} + \cos\underline{Q}\cdot\underline{r}_{23} + \cos\underline{Q}\cdot\underline{r}_{31} \ . \tag{4.17}$$

The scattering function, finally, reads

$$S_{inc}^R(\underline{Q},\omega) = \left[\frac{1}{3} + \frac{2}{9}A(\underline{Q})\right]\delta(\omega) + \left[\frac{2}{3} - \frac{2}{9}A(\underline{Q})\right]\frac{3w/2\pi}{\omega^2 + (3w/2)^2} \ . \tag{4.18}$$

By averaging over all orientations Ω_Q of the wave vector $\underline{Q}$ (3.19) the equivalent expression for a powder is obtained /4.4/

$$S_{inc}^R(Q,\omega) = \left[\frac{1}{3} + \frac{2}{3}j_0(Q\rho\sqrt{3})\right]\delta(\omega) + \left[\frac{2}{3} + \frac{2}{3}j_0(Q\rho\sqrt{3})\right]\frac{3w/2\pi}{\omega^2 + (3w/2)^2} \ . \tag{4.19}$$

If the scattering function does not refer to a single proton (as here) but to the molecule, there is an additional factor M_p (= number of protons in the molecule). Both the expressions for single crystals and powders fulfill the sum rule $\int S^R_{inc}(\underline{Q},\omega)\, d\omega = 1$ and $\lim_{Q\to 0} S^R_{inc}(\underline{Q},\omega) = \delta(\omega)$.

4.5 Reorientations in Ammonium Salts

Several experiments which have been analyzed with the rotational jump model are given in the review by LEADBETTER and LECHNER /1.10/. We will only consider the example of tetrahedral molecules and discuss two fairly recent measurements with ammonium salts /4.27,28/. The simplest model has been discussed by SKØLD /4.21/ and applied to the rotations in solid methane, though its application to ammonium salts seems more adequate. The reorientations of a XH_4 molecule at a tetrahedral site, which corresponds to four possible positions for each proton, is treated. Allowing only for a single type of jump (either 120° or 180° jumps), the quasielastic scattering is described by a single Lorentzian. With a restriction to jumps around the threefold axis one obtains for a powder sample

$$S^R_{inc}(Q,\omega) = \frac{1}{4}\left[1 + 3\, j_0(Q\rho\sqrt{3})\right]\delta(\omega) + \frac{3}{4}\left[1 - j_0(Q\rho\sqrt{3})\right]\frac{4u/\pi}{\omega^2+(4u)^2} \; . \qquad (4.20)$$

where u denotes the probability per unit time that a 120° jump occurs around a certain threefold axis. All required premises for this model are rather closely fulfilled in the orientationally ordered phase (CsCl structure, space group $P\overline{4}3m$) of the ammonium halides, well below the transition temperature into the ordered phase. In this temperature regime, unfortunately, the residence times are rather long and the quasielastic scattering cannot be resolved with presently available resolutions of neutron spectrometers.

4.5.1 $(NH_4)_2SnCl_6$

There is, however, another class of ammonium salts, which may serve as an example instead: $(NH_4)_2MX_6$ with M^{4+} = metal ion, X^- = halide ion. These salts mostly condense within the relatively simple antifluorite structure (space group Fm3m), which has tetrahedral symmetry ($\overline{4}3m$) at the ammonium site (Fig.4.7).

$(NH_4)_2SnCl_6$ is one of the hexahalometallates crystallizing within this structure /4.29/. The rotational potential is considerably weaker than in the ammonium halides,

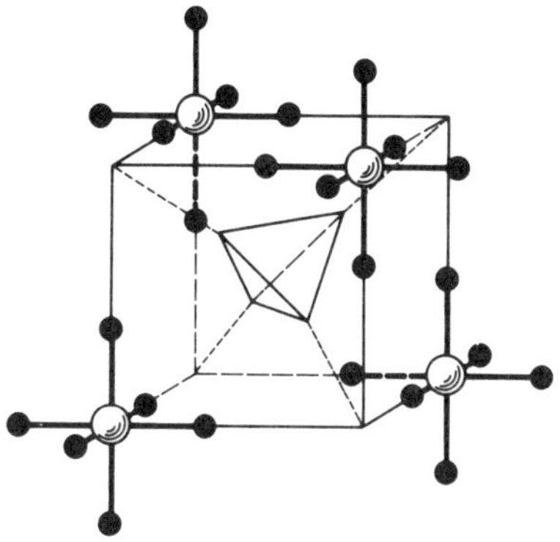

Fig.4.7. Structure of $(NH_4)_2SnCl_6$ (space group Fm3m). The figure emphasizes the tetrahedral coordination of the NH_4^+ group (position 1/4 1/4 1/4 by four $[SnCl_6]^{--}$ complexes at the face centers of the unit cell

thus allowing rotational jumps on a time-scale well accessible to neutron scattering. A quantitative description of the rotational potential on the basis of Coulomb interactions (which has been successful in the case of the perovskite NH_4ZnF_3 /2.23/ plus hard-core repulsion has been attempted but has not yielded a satisfactory result so far /4.27/. A structure analysis of the isostructural compound $(NH_4)_2SiF_6$ /4.30/ shows that the orientational probability function is not Gaussian and suggests that there is a wide potential well in which the NH_4^+ group librates. The situation is probably similar in $(NH_4)_2SnCl_6$. The existence of an extremely anharmonic, boxlike potential may be concluded from the energy of the excited librational states as $E_2 > 2E_1$ is observed /4.27,31/. Here E_1 and E_2 denote the energy of the first and second excited librational state, respectively.

Without detailed knowledge of the potential, it is difficult to estimate how much less important 180° jumps are in comparison with 120° jumps. This information could also be provided via the $\underset{\sim}{Q}$ dependence of the quasielastic scattering from a single crystal (Sect.4.5.2). Experiments with $(NH_4)_2SnCl_6$ so far have only been performed with powder samples and therefore the question, whether 180° jumps can be ignored or not still cannot safely be answered. In spite of that, $(NH_4)_2SnCl_6$ provides a very interesting system for the study of stochastic rotational motions. As there is no phase transition, rotational jumps can be observed in a rather wide temperature range (better: range of τ). It is limited at high temperatures ($T \cong 300$ K) by the fact that the jump model probably becomes inadequate and should be replaced by a model accounting for diffusion in a potential (Sect.4.6). The temperature range is limited on the low-temperature side as well because below $T \cong 70$ K quantum aspects become important (Sect.6.2.4a). Nevertheless the quasielastic linewidth Γ (HWHM) could be observed over almost three orders of magnitude (Figs.4.8,9). For this purpose two different spectrometers were employed: a three-axis spectrometer for the work with relatively coarse resolution and a backscattering spectrometer

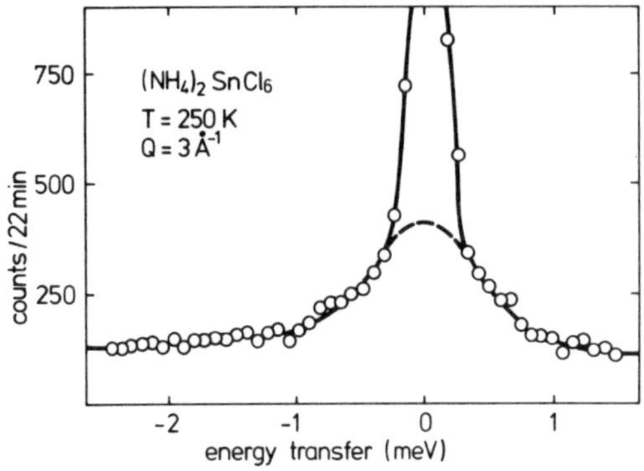

Fig.4.8

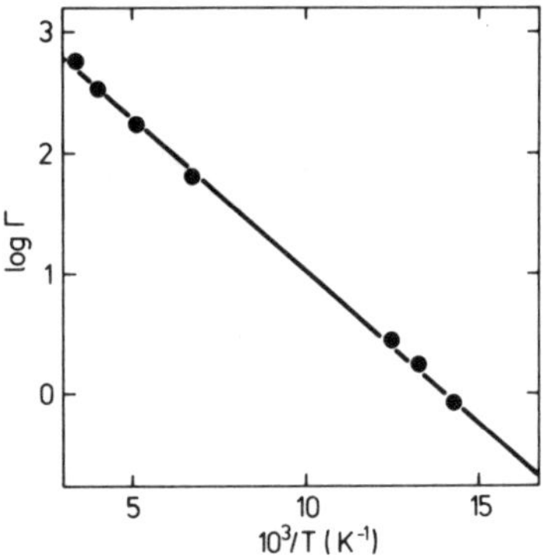

Fig.4.9

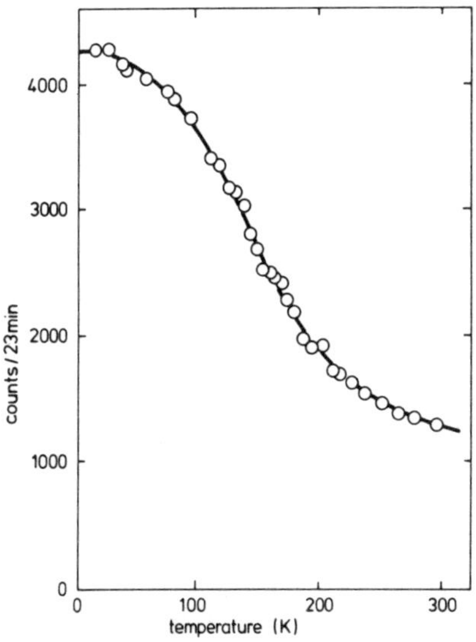

Fig.4.8. Quasielastic neutron scattering from $(NH_4)_2SnCl_6$ powder /4.27/. The halfwidth $\Gamma = 4u = 4\hbar/\tau$ of the quasielastic line is directly connected with the jump rate for classical reorientation

Fig.4.9. Arrhenius plot of the halfwidth $\Gamma(T)$ (in μeV); from the slope of the straight line fitted to the data point, an activation energy $E^A = 590 \pm 30$ K for the ammonium reorientation in $(NH_4)_2SnCl_6$ is found /4.27/

Fig.4.10. Fixed-window measurement of the quasielastic scattering from $(NH_4)_2SnCl_6$ /3.8/. A fit to the experimental data (solid line) yields an activation energy $E_A = 620 \pm 130$ K for the reorientational motion of the NH_4^+ ions

for the high-resolution spectra. The temperature-dependent background which has been noted in the spectra recorded with coarse resolution and which could not be explained in /4.27/, probably is due to acoustic phonons /3.8/. The spectra are analyzed by least-squares fits on the basis of a simple model. The quasielastic scattering is described as a Lorentzian with temperature-dependent width $\Gamma(T)$ (4.20), which is convoluted with the instrumental energy resolution.

Figure 4.9, in which log $\Gamma(T)$ is plotted versus 1/T, exhibits an Arrhenius type of behavior. From the slope of the straight line through the data points, an activation energy $E^A = 590 \pm 30$ K is deduced /4.27/, which is in rather good agreement with NMR values. Obviously thermally activated jumps across the potential barrier take place and yield $\Gamma = \Gamma_0 \exp(-E^A/kT)$ with $\Gamma = 4u$ [the jump rate u is defined in connection with (4.20)].

The activation energy E^A can be determined in a different way which avoids the temperature-dependent measurement of quasielastic spectra: the "fixed window" method /4.32/. The spectrometer is set to the elastic position, that is, to an energy transfer $\hbar\omega = 0$ and the finite energy resolution of the spectrometer defines an "energy" window. The intensity in the counter I_{FW} strongly depends on the linewidth $\Gamma(T)$, which changes with temperature.

$$I_{FW}(\underline{Q},T) \sim \exp(-2W) \int_{-\infty}^{\infty} R(\omega)\, S^R_{inc}(\underline{Q},\omega)\, d\omega \quad . \tag{4.21}$$

Here $R(\omega)$ denotes the energy resolution of the instrument at a nominal energy transfer $\hbar\omega = 0$. A fixed window scan /3.8/ with $(NH_4)_2SnCl_6$ and an energy resolution $\Delta E \cong 1$ meV (FWHM) is shown in Fig.4.10. It yields an activation energy $E^A = 620 \pm 130$ K. As to be expected the uncertainty of E^A is larger than with the measurement of quasielastic spectra. The advantage of the fixed window technique is its simplicity which renders the method extremely useful for a first survey in an investigation of a new system. Its usefulness was demonstrated convincingly in a measurement with $Pb(CH_3)_4$ /4.32/. Leadtetramethyl is a tetrahedral molecule with methyl groups at the corners of the tetrahedron. Two activation steps were found (Fig.4.11) and attributed to the stochastic rotational motion of the molecule as a whole with a relatively large activation barrier $E^A_1/k = 190$ K and CH_3 group rotation with a much smaller activation barrier $E^A_2/k = 7$ K. A value $E^A \sim B$ (rotational constant) should not be taken too literally, however, as the activation step probably extends beyond the regime of classical rotation.

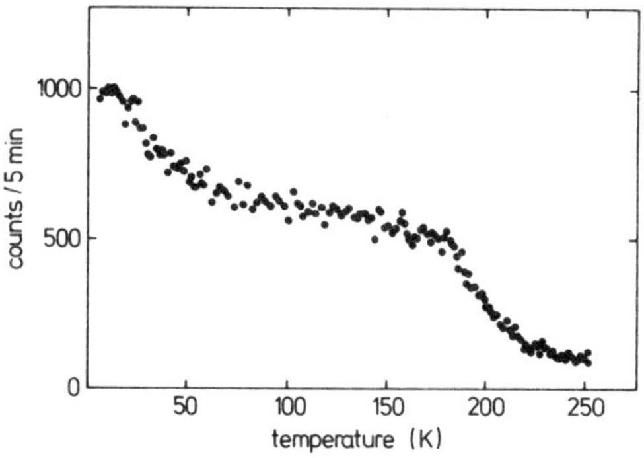

Fig.4.11. Fixed-window measurement of the quasielastic scattering from $Pb(CH_3)_4$. Two activation steps are observed and yield activation energies $E^A_1 = 190$ K and $E^A_2 = 7$ K, which have been related with the rotational motion of the whole molecule and that of single methyl groups, respectively /4.32/

4.5.2 Ammonium Chloride (NH_4Cl)

A second example of considerable interest is provided by the ammonium halides, particularly by NH_4Cl in the vicinity of its orientational order-disorder transition at 242 K /4.33/. Above the transition temperature the ammonium sites have symmetry m3m which is greater than the tetrahedral symmetry of the NH_4^+ group. Consequently the molecules have two distinguishable orientations or—more precisely—there are 2 sets of 12 orientations (described by 12 different Euler angles ω^E), each orientation within a given set generating the same density distribution (Fig.4.12). Below the phase transition one of the orientations with occupation p_1—we return to the simplified picture of just two orientations—becomes preferred. In the ordered phase with the tetrahedra aligned parallel, one may introduce an order parameter $\eta = p_1 - p_2$. Above T_c in the disordered phase the order parameter is zero, as $p_1 = p_2 = 1/2$.

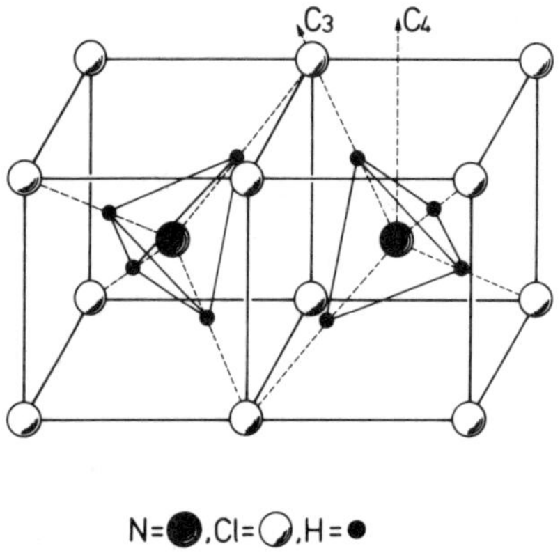

Fig.4.12. Structure of NH_4Cl (space group Pm3m), with the two alternative orientations of the NH_4^+ ion. C_3 and C_4 denote axes for 120° jumps and 90° jumps, respectively /4.28/

MICHEL /4.34/ has discussed the reorientational motion above T_c, where each proton may be found in one of eight possible positions. The rotational jump model seems to be very appropriate for stochastic rotational motions in NH_4Cl, as the molecules are strongly localized with respect to their orientations; the librational amplitude is only about 7°. Two different kinds of jumps are distinguished: 120° jumps around the threefold axis (rate u) which do not affect the order parameter and 90° jumps (rate v) which change the orientation of a molecule and thus the order parameter η. For the disordered phase MICHEL /4.34/ finds that apart from the totally symmetric solution with $\lambda_1 = 0$, there are 3 eigenvalues connected with the quasielastic scattering: $\lambda_2 = 2v + 4u$ and $\lambda_4 = 6v$ are triply degenerate eigenvalues, $\lambda_3 = 4v + 4u$ is not degenerate. Each Lorentzian of width λ_i is connected with a specific $\underline{Q}$-dependent form-factor /4.34/. Obviously an attempt to determine u(T) and v(T) separately — in order to test Michels's prediction of a strong change of v in the vicinity of the phase transition — requires $\underline{Q}$-dependent measurements in a single crystal. Such a measurement has recently been performed by TÖPLER et al. /4.28/. It has been performed in an extremely careful fashion, rendering it a model example for reorientation studies. Below the phase transition the cubic potential becomes distorted, a ratio $p_2/p_1 = \exp(-\Delta E_4^A/kT) \neq 1$ results and consequently, the 90° jumps from (rate v_1) or to (rate v_2) the favorable positions become inequivalent. ΔE_4^A is the difference of the ground-state energies at the two sites. Therefore, in order to analyze data taken at $T < T_c$, it turned out to be necessary to solve the rate equations for a nonzero order parameter η /4.28/. A calculation within mean-field approximation (more correctly, the λ_i would depend on the nearest-neighbor correlation function) yields generalized eigenvalues λ_i $(i = 2,3,4)$

$$\lambda_2 = 4u + \frac{v_1}{1-\eta}\left(3 - \sqrt{1+8\eta^2}\right) \tag{4.22a}$$

$$\lambda_3 = 4u + \frac{v_1}{1-\eta}\left(3 + \sqrt{1+8\eta^2}\right) \tag{4.22b}$$

$$\lambda_4 = \frac{6v_1}{1-\eta} \quad . \tag{4.22c}$$

Here the relation $v_2 = v_1(1+\eta)/(1-\eta)$ has been used. Both limiting cases are reproduced. For $\eta = 0$ MICHELS's original result /4.34/ is found, while $\eta = 1$ leads to the expression obtained by SKØLD /4.21/. For simplicity the authors have not distinguished between two different types of 120° jump rates (u_i), corresponding to the two nonequivalent orientations. The results, as obtained with thin single crystals of NH_4Cl (transmission about 80%) with high-resolution neutron spectroscopy, are shown in Fig.4.13. They are based on measurements at several $\underline{Q}$ positions, chosen such as to single out either λ_2 or λ_4, to the extent possible. Only if $\eta(T)$ in the ordered phase is known, can the jump rates u and v (120° and 90° degree jumps, respectively) be extracted from the quasielastic linewidth $\Gamma(\underline{Q},T)$. η has been determined from measurements of the Bragg intensities on a three-axis spectrometer. This also allowed the determination of the Debye-Waller factor which is needed for a quantitative analysis of $S_{inc}(\underline{Q},\omega) = e^{-2W} S_{inc}^R(\underline{Q},\omega)$.

Figure 4.13 shows that the 90° jump rate v first decreases continuously on approaching T_c from above, but then drops sharply at the transition temperature. This is due to the first-order character of the transition, which ultimately leads to a discontinuous change of η as well as ΔE_4^A from zero to a finite value. To the extent, the second orientation (probability p_2) becomes forbidden, 90° jumps also become less probable. The transition, by the way, becomes continuous at pressures above

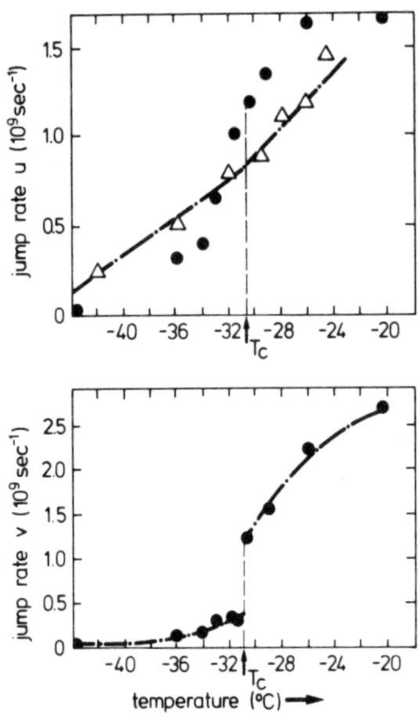

Fig.4.13. The upper part of the figure represents results for 120° jumps (rate u) in NH_4Cl; different symbols denote two different measurements (more precise measurement: triangles). Corresponding results for 90° jumps (rate v) are shown in the lower part of the figure /4.28/

the tricritical point at $p_T \cong 1.5$ Kbar. The 120° jump rate u also decreases, but without a noticeable jump at T_c /4.28/.

There are several other aspects, which are worth noting. With use of the relation $v = v^0 \exp(-E_4^A/kT)$ and the semi-quantum-mechanical prefactor $v^0 = (kT/h)\cdot[1-\exp(-E_1/kT)]$ /4.35/, $E_4^A/k = 1985$ K can be deduced for $T = T_c + 1$ K and $E_4^A/k = 2260$ K for $T = T_c - 1$ K. The energy of the first librational state is $E_1 = 49$ meV /4.36/. Classically $v^0 = E_1$ is an attempt frequency. The activation energy E_4^A which is obtained by the above method /4.28/ is close to, but smaller than the value derived from the rotational potential $V(\tau)$ of HÜLLER and KANE /2.16/ via $E_4^A = V(\tau_b)-V(\tau_m) - \frac{1}{2}\hbar\omega_{LIB}$; $(\hbar\omega_{LIB} = E_1)$. Here τ_m and τ_b denote the orientations at the minimum and between two minima (passage from the minimum to the other by fourfold rotation), respectively. The difference probably is due to the fact that there is a saddle point close to τ_b, but not at τ_b. Presently molecular dynamics calculations are performed /4.37/ with the aim of studying details of the passage from one minimum to another.

4.6 Diffusion in the Presence of a Potential

Before leaving the field of classical molecular rotations, we want to discuss the situation prevailing in most molecular crystals—the diffusion in the presence of a finite rotational potential. So far there are no neutron experiments available which are aimed at testing recent model calculations. Therefore we shall restrict ourselves to a brief description of two alternative approaches.

DIANOUX and VOLINO /4.38/ have calculated the scattering functions $S^R_{inc}(\underline{Q},\omega)$ starting from a diffusion equation for $G_s(\underline{r},t)$ in presence of a rotational potential $V(\phi) = \frac{1}{2} V_n \cos n\phi$. The replacement of the Langevin equation by a diffusion equation for the self-correlation function $G_s(\underline{r},t)$ apparently is a reasonable approximation in the strong friction limit (Sect.4.2). Otherwise the short-time behavior of $G_s(\underline{r},t)$ which is not adequately described within the above approximation, becomes important. The model of Dianoux and Volino correctly reproduces continuous rotational diffusion and a rotational jump model between n equidistant orientations as limiting cases. In the intermediate regime no analytic expression can be given as the solutions are obtained by truncating an otherwise infinite matrix before it is diagonalized. Though the method has not yet been used to analyze rotations in a specific molecular crystal, the approach means a significant step forward. It may be hoped that a generalization to three-dimensional diffusion soon will follow. Other approaches which might favorably be translated into the language of rotational motion have been performed in context with ionic conduction. There the translational diffusion of a particle in a periodic potential has recently been studied /4.39,40/.

Another promising approach for the treatment of classical rotations in a periodic potential recently has been published by de RAEDT and MICHEL /4.41,42/. The authors presented a model which combines an oscillatory aspect—at low temperatures the molecules perform angular oscillations around one of their equilibrium orientations—and a stochastic aspect which may be described by thermally activated reorientations. Obviously the oscillatory aspect dominates at low temperatures, where the reorientational motion is exceedingly slow. Because of the classical treatment, the model is not applicable at very low temperatures, however. At high temperatures the reorientations become increasingly important. The authors used symmetric-adapted surface harmonics $K_{\ell m}[\Omega(t)]$ as dynamical variables and a continued fraction approach (which will not be described here) in order to calculate the dynamics of a single molecule in an effective potential. The starting point is the single particle Hamiltonian $\mathcal{H}$ of the molecule (see Sect.5.1). For calculating the time-dependent correlation functions $\langle K_{\ell m}[\Omega(t)]\ K_{\ell m}[\Omega(t=0)]\rangle$ the time evolution of the dynamical variables $K_{\ell m}[\Omega(t)]$ is needed and classically can be obtained via the Poisson brackets $\dot{K}(t) = \{K(t),\mathcal{H}\}$. So far the approach is restricted to molecular impurities in crystals. The explicit calculation /4.41/ is specialized to the motion of a dumbbell in an octahedral cage (e.g., CN^- in KCl; see Fig.2.1). A formulation of both the incoherent neutron scattering law and of the Raman intensities is given. The relation of the energy width of the quasielastic peak (reorientation) and that of the inelastic peaks (oscillation) to the widths generated within the Langevin model of diffusion is not yet established.

As there are practically no examples for the classical rotation of dumbbell-type molecules with a large σ_{inc}, a calculation aiming at tetrahedral molecules or methyl groups is highly desirable. It is for these reasons that the calculation in its present form /4.41/ is compared to Raman measurements with CN^- impurities in alkalihalides /4.43/. Very good qualitative agreement is found.

An interesting aspect concerns the character of the rotational motion for dynamical variables of different symmetry. Spherical harmonics of order $\ell = 2$ can be decomposed into two harmonics of E_g symmetry and three harmonics within T_{2g}. For a potential $V(\Omega) = V_4 K_{41}(\Omega)$ (2.19) and V_4 positive, the potential minima are in [111] and the maxima in [100]. For a negative sign of V_4 this is reversed. The authors show that harmonics of E_g symmetry have reorientational character for $V_4 > 0$, and the harmonics of T_{2g} symmetry have oscillatory character. For $V_4 < 0$ the situation is reversed. The reorientation between three equivalent orientations (minima in [100]) is isomorphous to the case of methyl group rotations. There E symmetry is found for the relaxation of the probability densities $p_j^{(i)}$ (Sect.4.4). Similarly the symmetry character of the angular oscillations can be shown to be that found in /4.41/.

5. Rotational Excitations at Low Temperatures
I. Principles

At low temperatures the fluctuations of the rotational potential die out and quantum aspects determine the single particle rotation. When fluctuations can be neglected it suffices to solve a stationary Schrödinger equation in order to learn about energy eigenvalues and eigenfunctions of a molecule in a given rotational potential. Direct measurements of the transitions between rotational states in general require neutrons. Transitions within the ground-state multiplet usually are accompanied by a change of the nuclear-spin function which can be produced by neutrons. This aspect will be dealt with somewhat later. Measurements with molecules displaying almost free rotation in a quantum-mechanical sense have been known for quite a while — particularly the transition from p-H_2 to o-H_2 which involves a change of the rotational quantum number J from an even to an odd value (e.g., $J=0\rightarrow 1$) /5.1,2/. Another example is solid methane (CH_4) in its phase II; there 2 out of 8 molecules (Fig.6.1) remain disordered at low temperatures and rotate almost freely /5.3-5/. Rotational tunneling only recently has opened up as an active field for neutron scattering /5.5/ and has stimulated a lot of activity, both experimental and theoretical. This does not mean that almost free rotation is the more frequently found phenomenon — on the contrary. Due to intermolecular interactions in a solid, only in rare cases is V/B a small quantity. Rotational tunneling has not been observed earlier because high resolution neutron spectrometers ($\Delta E \gtrsim 0.3$ µeV or 4 mK) had not been available before about 1973.

The natural unit of the periodic rotational potential is the rotational constant $B = \hbar^2 2\Theta$; examples are $B(H_2) = 85$ K and $B(NH_3, CH_3, NH_4^+, CH_4)$ all about 7-8 K. For all other molecules B is of the order of 0.1 K or less, e.g., $B(CCl_4) \sim 0.08$ K. Obviously hydrogen is close to the limit of quantum-mechanical free rotation, especially because the anisotropic interaction (o-H_2 has a quadropole moment) is only of the order of a few K. For large molecules with heavy constituents — the above example is CCl_4 — the rotational constant is very small. As in addition the interaction is strong, the rotational motion at low temperatures is quenched. Low-energy rotational excitations accessible to neutrons are found in CH_4, NH_4^+ salts, NH_3 and CH_3 groups in various surroundings and crystal phases.

In this section the calculation of rotational states is described starting from the single molecule Hamiltonian. Two alternative approaches have been used. Dealing with homonuclear molecules the exclusion principle enters and properly symmetrized wave functions have to be constructed as sums of products of rotational wave functions Φ and spin functions χ. Only after the construction of complete wave functions is the basis for calculating neutron scattering transition matrix elements provided.

In connection with rotational tunneling, a plausible explanation for its single particle character can be given. While the wave functions Φ of two states with different symmetry differ appreciably, the densities $\Phi\Phi^*$ (or the corresponding charge density) are almost identical. Therefore, the intermolecular interaction is practically unaffected by a tunneling transition.

5.1 Rotational States

5.1.1 One-Dimensional Rotation and Exact Methods

We will start the discussion of the quantum-mechanical states of a molecule in a periodic potential by giving an especially simple example, that of a dumbbell which is restricted to only one angular degree of freedom. With a site symmetry which is a subgroup of (or equal to) the molecular symmetry the leading term in the potential is $V(\phi) = (1/2)V_2\cos 2\phi$. The kinetic energy is $B\ \partial^2/\partial\phi^2$ and the following Schrödinger equation is obtained

$$\left(-B\frac{\partial^2}{\partial\phi^2} + \frac{1}{2}V_2\cos 2\phi\right)\psi = E\psi \quad . \tag{5.1}$$

A simple substitution, $a = E/B$ and $q = V_2/4\ B$, leads to the well-known Mathieu equation /5.6/ which is exactly soluble. Expressions for the eigenvalues E_i and the eigenfunctions ce_i and se_i are given in /5.6/. Usually the E_i are expressed in terms of continued fractions /5.7,8/. For details of the derivation, the reader is referred to the papers of GLODEN /5.8/ who also has adapted the method to general potentials of the shape $(V_n/2)\cos n\phi$. This allows the calculation of eigenstates for various kinds of uniaxial rotors including those of CH_3 groups in potentials of threefold or sixfold symmetry /5.8/. The solutions of the Mathieu equation also are extremely useful for testing approximations, which are necessary in case of more than one angular degree of freedom.

The lower energy eigenvalues of the Mathieu equation with a term $q\cos 2\phi$ are shown in Fig.5.1. Three different regimes may be distinguished.

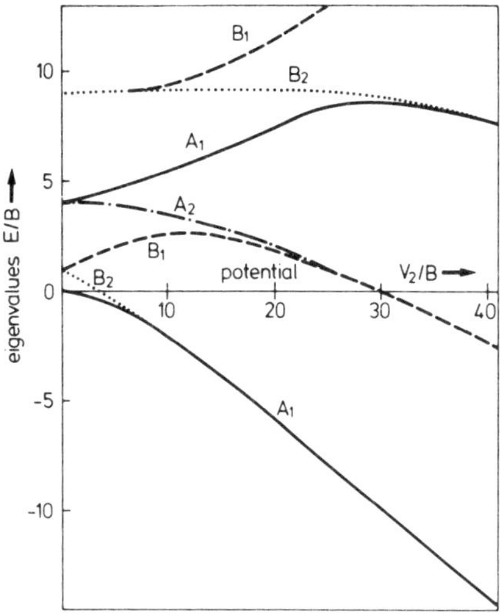

Fig.5.1. Energy eigenvalues of a dumbbell with one rotational degree of freedom in a potential $V(\phi) = 1/2\ V_2\cos 2\phi$ (solution of the Mathieu equation)[2]

a) *Weak rotational potential.* For $V_2 = 0$ one retains a quantum-mechanically free rotor with energy eigenvalues $E_J = BJ^2$ and eigenfunctions $ce_J(\phi;V_2=0) = \cos J\phi$ and $se_J(\phi;V_2=0) = \sin J\phi$ (norm omitted). All states are labelled by the rotational quantum number J; only the ground state with J = 0 is nondegenerate—all excited states ($J \geqq 1$) are doubly degenerate. For small V_2 the potential can be treated as a perturbation and J still can be taken as a good quantum number. For comparison we also give the energies E_J and degeneracies M_J (of the spatial wave functions) for two- and three-dimensional rotors: 1) linear molecule: $E_J = BJ(J+1)$, $M_J = 2J+1$; 2) spherical top molecule: $E_J = BJ(J+1)$, $M_J = (2J+1)^2$.

b) *Strong rotational potential.* For $V_2 \gtrsim 30$ B the molecules perform almost harmonic librations (for more angular degrees of freedom this limit is approached for somewhat higher values of V_2). In the limit of large V_2 the energy eigenvalues are $E_n = -V_2/2 + (n+1/2)\frac{2}{\hbar}V_2 \cdot B$ and the corresponding functions are oscillator functions. As may be found in the expression for E_n, the separation between the oscillator levels increases with increased orientational localization, i.e., increasing strength of the potential. The degeneracy of all states is two in the example chosen.

c) *Intermediate regime.* In the intermediate regime the eigenstates are neither oscillator functions nor free rotor functions. The states are characterized by their symmetry and transform according to the representations of the group C_{2V} (Table 5.1). The four symmetry operations, described in Table 5.1, leave the potential $V(\phi) \sim \cos 2\phi$ invariant. The ground-state wave function $ce_0(\phi)$, for example, has symmetry A_1 and therefore may be expressed in terms of functions $\cos 2n\phi$ only. Of particular interest is the regime close to the limit of librational motion. Here the states look

Table 5.1. Character table of the point group C_{2v}(mm); listed are also the angular transformations with correspond to a given symmetry operation and the trigonometric functions which transform according to the four one-dimensional representations of C_{2v}

C_{2v}	E	C_2	$\sigma(x)$	$\sigma(y)$	
transform. of angular variable	ϕ	$\pi+\phi$	$2\pi-\phi$	$\pi-\phi$	
A_1	1	1	1	1	$\cos 2n\phi$
A_2	1	1	-1	-1	$\sin 2n\phi$
B_1	1	-1	1	-1	$\cos(2n+1)\phi$
B_2	1	-1	-1	1	$\sin(2n+1)\phi$

almost like harmonic-oscillator states. Closer inspection, however, reveals that the eigenstates are split in two closely spaced levels of different symmetry [groundstate: A_1 for $ce_0(\phi)$ and B_2 for $se_1(\phi)$]. Within the framework of pocket states, that is, wave functions centred at the minima of a given rotational potential (potential pocket), the splitting is caused by the overlap of wave functions located in adjacent potential wells. A schematic drawing of tunnel split rotational states in a periodic potential is shown in Fig. 5.2 (it refers to the more realistic situation of a $\cos 3\phi$ potential). An example of pocket states in a $\cos 6\phi$ potential is shown in Fig.5.3. The solutions of the Mathieu equation also provide a "correlation diagram", that is, they show how the oscillator states emerge from the free rotor states when the potential is increased. Obviously the multiplicity M_0 of the combined groundstate levels can be identified with the total number of potential pockets.

As mentioned above the calculation of eigenvalues has not been restricted to a uniaxial rotation of dumbbells, but mainly concentrated on rotating XH_3 groups (X = C,N). Both the molecular symmetry and the site symmetry have to be included. A careful investigation of this aspect is given by KING and HORNIG /5.9/ for three-dimensional rotors. For CH_3 groups the symmetry arguments are much simpler; the molecular symmetry causes that only Fourier components $V_n \cos(n\phi+\phi_n)$ of the potential $n = 3m$ ($m = 1,2,3...$) matter. The site symmetry may help to further reduce the nonvanishing terms in the potential and to impose conditions on the phase angles ϕ_n. For example, a twofold axis in the crystal along the symmetry axis of the molecule eliminates contributions other than V_{6m} and a mirror plane containing the symmetry axis of the molecule will render $\phi_n \equiv 0$ (see also Sect.4.3.1 and Fig.2.2).

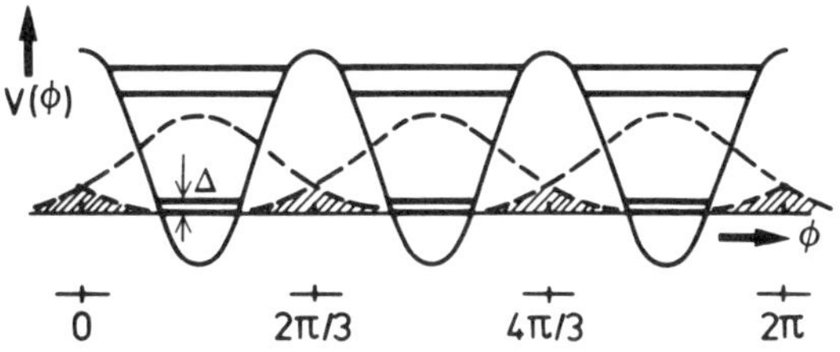

Fig.5.2. Schematic drawing representing the tunnel splitting of the rotational ground state (Δ) and of the first librational state. The splitting is due to the overlap of wave functions (dashed region) in neighboring potential wells

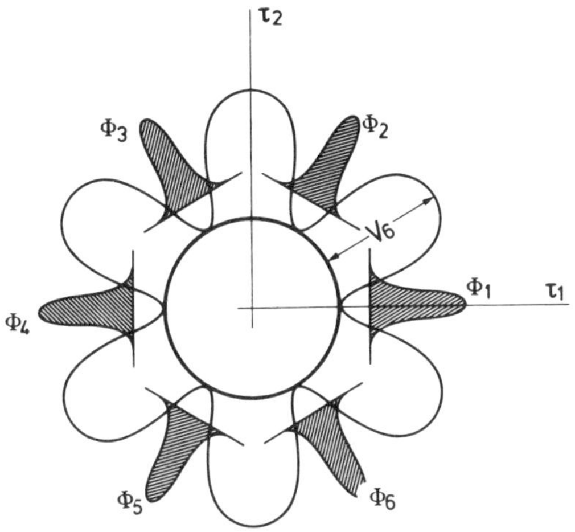

Fig.5.3. Schematic drawing of a one-dimensional rotational potential with a sixfold symmetry axis. Also shown are the pocket states Φ_i /5.8/. The drawing might represent a disordered CH_3 group (sixfold symmetry axis at the molecular site)

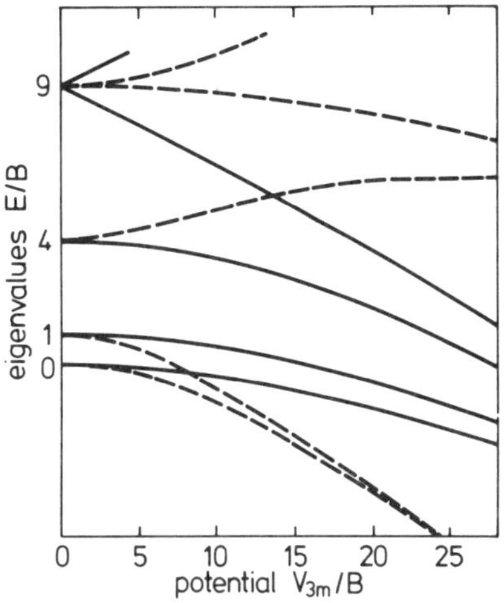

Fig.5.4. Energy eigenvalues of a XH_3 group in potentials $V(\phi) = (1/2)V_{3m}\cos 3m\phi$ /5.8/. The dashed line refers to a $\cos 3\phi$ potential, the solid line to a $\cos 6\phi$ potential

Solutions of the Schrödinger equation of a one-dimensional rotor have been given for potentials of the form

$$V(\phi) = \frac{1}{2} V_3 \cos 3\phi + (-1)^k \frac{1}{2} V_6 \cos 6\phi \tag{5.2}$$

with k = ±1 and are tabulated for several sets of parameters V_3 and V_6 /5.8/. The results for either $V_6 \equiv 0$ or $V_3 \equiv 0$ are shown in Fig. 5.4. Some of the eigenvalues may be derived directly from the results obtained from (5.1).

5.1.2 Approximate Methods

Approximate methods have only littly immediate importance for the case of uniaxial rotation. They become necessary for rotation with two and three angular degrees of freedom. Nevertheless, approximate methods have been used also in the context of uniaxial rotation /5.10,11/. As has been pointed out by HÜLLER and KROLL /5.9/, the comparison with exact solutions (which are available for the one-dimensional problem) allows a direct test of the quality of a given approximation.

There are two different approaches which appear possible and have been used mainly for two- and three-dimensional rotation. 1) One method uses an expansion into free rotor wave functions. 2) The other employs states which resemble harmonic oscillator functions. In the first case one is lead to the diagonalization of infinite Hamiltonian matrices $<\Phi_\ell|\mathcal{H}|\Phi_{\ell'}>$. The problem is somewhat simplified by using symmetry-adapted wave functions (SAF); the indices ℓ, ℓ' then include both angular momentum and symmetry labels. The use of SAF's leads to block-diagonal matrices. Nevertheless, it is necessary to truncate the matrix by restricting it to a subspace $J \leq J_0$ of the angular momentum quantum number J.

The method undoubtedly works well for weak potentials, where rapid convergence of the expansion of the wave function is guaranteed. Problems, particularly for three-dimensional rotators, arise in the strong potential limit. There rather large cutoff values J_0 are required for sufficient accuracy in the determination of the eigenvalues. This, however, considerably blows up the dimension of the truncated matrices—in particular for the eigenvalue problem with SAF's of low symmetry (e.g., $V(\omega^E)$ with cubic symmetry; for representations with T symmetry the dimension is ≥ 200, with $J_0 \cong 13$ /5.12/).

The other way is the use of harmonic-oscillator-like wave functions. One starts with pocket states Φ_i (as introduced before) which do not represent eigenstates of the Hamiltonian. If prepared in one of the potential pockets, a state is not stationary, but decays with time into other pocket states because of the finite probability of tunneling. The magnitude of the tunnel splitting is determined by overlap matrix elements $H_{ij} = <\Phi_i|\mathcal{H}|\Phi_j>$; here i and j denote different pockets. For the geometry sketched in Fig.5.3 there are three independent matrix elements H_{12}, H_{13}, and H_{14}, for example. As in the case of the expansion into free rotor functions, symmetry helps to block-diagonalize the Hamiltonian matrix. The method has been demonstrated in an especially clear fashion by HÜLLER and KROLL /5.13/, who used Gaussian-shaped pocket states, e.g., $\Phi_i = \exp(x_G\tau_i^2)$. The quantity x_G is taken as a variational parameter. For strong potentials x_G is large and Φ_i becomes a narrow Gaussian. The width $x_G^{-1/2}$ is determined by minimizing the eigenvalues after the diagonalization of the Hamiltonian matrix. It depends on both the magnitude of the potential and — to a lesser extent — the symmetry of the state. The method leads to exact answers in the limit of weak and strong potentials. In the latter case, however, this is only true for the librational energies and not for the tunnel splittings. The calculated splittings for a potential $V(\phi) = (1/2)\, V_6 \cos 6\phi$ turn out to be about half of the exact value /5.13/ for $V_6 \cong 150$ B (Fig.5.5). The discrepancy becomes even worse for larger V_6.

One immediately suspects that this results from the fact that Gaussians (with just one parameter for the width) do not well represent the wave functions in the overlap region, although they are a very good approximation in the region of the potential minimum. The magnitude of the wave function in the overlap region, however, is decisive for the overlap matrix elements. Therefore improved pocket state functions are needed and a step in this direction recently has been made /5.14/. There the Gaussian is multiplied by the first few terms of a symmetry-allowed polynomial within the angular coordinates, which may be understood as an admixture of excited harmonic-oscillator states to the ground-state function. This also allows for a spatial anisotropy of the wave function in the region of the minimum and certainly is a better representation in the overlap region. With improved pocket states an increased overlap results and, consequently, larger tunnel splittings are obtained /5.14/. In /5.14/ the variation principle is applied to wave function with several parameters.

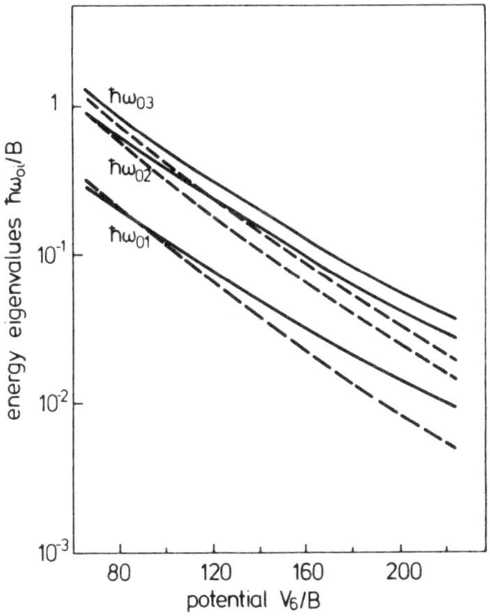

Fig.5.5. Ground-state tunneling frequencies of a one-dimensional rotor in a potential $V(\phi) = (1/2)\, V_6 \cos 6\phi$. ω_{0i} refers to transitions from J = 0 to J = 1, 2 and 3, respectively (perturbed by potential). Solid curve: exact results; dashed curve: results of a variational calculation using pocket states. Both potential and energy eigenvalues are given in units of the rotational constant $B = \hbar/2\Theta$ /5.13/

An advantage of the pocket state formalism is the very small set of basis functions which is required. On the other hand its application is difficult in connection with problems with many potential minima, such as in the case of CH_4 as a substitutional impurity in rare gas matrices /5.15/. We will return to the comparison of the two techniques in connection with the rotational states of a tetrahedron.

5.1.3 Two-Dimensional Rotation (Linear Molecules)

Rotational tunneling of linear molecules in stoichiometric crystals so far has not been observed by incoherent neutron scattering. Practically all experiments deal with the rotation of ionic molecular groups as impurities in alkali halides, and employed techniques other than neutron scattering. Much of this is discussed in reviews by NARAYANAMURTI and POHL /5.16/ on tunneling states of defects in solids and also by BARKER and SIEVEN /5.17/. In one case coherent neutron scattering has been successfully applied, namely with 0.1% CN^- in KCl /5.18/. There the rotational excitations were not observed directly, but via their coupling to acoustic phonons of the same symmetry. Dispersion curves with the same symmetry label cannot cross ("anticrossing" or "level repulsion"). Therefore the rotational mode acquires more and more acoustic character in the region of anticrossing. The inelastic structure factor of the acoustic modes is more favorable for an observation which renders the experiment possible. An explanation of the unexpectedly pronounced anticrossing has been reported in /5.19/. An attempt to observe the same phenomenon in solid argon with N_2 molecules as substitutional impurities failed /5.20/.

As most experiments have been done with cubic systems, theoretical efforts at calculating the rotational excitations of linear molecules also concentrated on sites with cubic symmetry. First calculations have been performed by DEVONSHIRE /2.29/, who retained only the first term $a_{41}K_{41}(\Omega)$ in an expansion of the potential into cubic harmonics. Free rotor wave functions have been used with a truncation at $J = 7$. Later authors /5.21-24/ extended this subspace to large J (/5.21/, $J \leq 27$), still employing the Devonshire potential. The results depend not only on the magnitude but also on the sign of the potential. For a positive sign the 8 potential minima are along [111] directions, while the 6 minima are along [100]; This is reversed for a negative sign. A localization along [110] always indicates the presence of higher order terms in the potential. An extension to a potential $V(\Omega) = a_{41}K_{41}(\Omega) + a_{61}K_{61}(\Omega)$ has been given by BEYELER /5.23/. His calculation uses a subspace $J \leq 12$ and is presented in the form of diagrams for various combinations of a_{41} and a_{61}. He also quoted the rotational states in a potential $V(\Omega) = a_{81}K_{81}(\Omega)$. Results for the rotational states of N_2 molecules in β-N_2 have been given by DUNMORE /2.26/. His paper contains a careful consideration of the rotational potential in

β-N_2 on the basis of intermolecular interactions and provides energy eigenvalues in a potential of hexagonal symmetry ($\overline{6}2m$). It appears, however, that the assumption of stationary rotational wave functions in the disordered phase of nitrogen is incorrect and that fluctuations of the potential are important.

5.1.4 Three-Dimensional Rotation (Tetrahedral Molecules)

When talking about rotational tunneling of molecules with a single rotational degree of freedom one essentially talks about CH_3 side groups and NH_3 molecules. In a similar way all work on rotational tunneling of three-dimensional rotors so far concentrated on tunneling states of tetrahedral molecules such as methane and the ammonium ion in a number of its salts. Theoretical approaches also concentrate on tetrahedral molecules. The ground-state splitting was first considered in connection with the zero point entropy of isotopic methanes by NAGAMIYA /5.25/. The method described there essentially is a pocket state approach; however, only symmetry arguments are used to distinguish between the different overlap matrix elements $<\Phi(\hat{R})|\mathcal{H}|\Phi(\hat{E})>$. Here $\hat{R}$ denotes one of the 12 symmetry elements of the tetrahedral group T (= 23 = subgroup of proper rotations of point group $\overline{4}3m$); $\hat{E}$ is the identity operator and $\Phi(\hat{R})$ is a shorthand notation for $\hat{R}\Phi$. In absence of any site symmetry other than the identity there are three independent 180° overlap matrix elements H_{x_i} and four pairs of 120° overlap elements h_i. The complete Hamiltonian matrix as set up by HÜLLER /3.3/ is shown in Table 5.2. NAGAMIYA solved the eigenvalue problem for trigonal and tetrahedral site symmetry /5.25/. Symmetry reduces the number of independent matrix elements. For tetrahedral site symmetry there is just one 120° overlap matrix element $h = h_1 = h_2 = h_3 = h_4$ and, similarly, just one 180° overlap matrix element $H = H_x = H_y = H_z$ (Fig.5.6). Diagonalization yields states with A, E and T symmetry with the following eigenvalues

$$
\begin{aligned}
E_A &= D + 3H + 8h \quad &&\text{(singlet)}\\
E_T &= D - H \quad &&\text{(three triplets)}\\
E_E &= D + 3H - 4h \quad &&\text{(doublet)}
\end{aligned}
\qquad (5.3)
$$

with $D = <\Phi|\mathcal{H}|\Phi>$.

The method sketched by NAGAMIYA /5.25/ has been considerably refined by HÜLLER and KROLL /5.13/ and HÜLLER /3.3/. In the latter publication the unitary matrix which block-diagonalizes the Hamiltonian matrix is given explicitly as well as the resulting 3×3 blocks connected with the T states (Table 5.3). There are three identical blocks, each with eigenvalues E_{T_1}, E_{T_2} and E_{T_3}. For low symmetry all three

Table 5.2. Hamiltonian matrix H_{ij} for the pocket states Φ_i of a tetrahedral molecule. D is the diagonal matrix element. H_x, H_y and H_z are the overlap matrix elements for 180° rotations around the x, y, and z axes, respectively. The matrix elements for 120° rotations around the 1, 2, 3 and 4 axes, respectively, are denoted by h_1, h_2, h_3 and h_4. The rotation axes which are defined in Fig.5.5 are fixed in the crystal frame /3.3/

D	H_x	H_y	H_z	h_4	h_3	h_1	h_2	h_4	h_2	h_3	h_1
H_x	D	H_z	H_y	h_2	h_1	h_3	h_4	h_3	h_1	h_4	h_2
H_y	H_z	D	H_x	h_3	h_4	h_2	h_1	h_1	h_3	h_2	h_4
H_z	H_y	H_x	D	h_1	h_2	h_4	h_3	h_2	h_4	h_1	h_3
h_4	h_2	h_3	h_1	D	H_x	H_y	H_z	h_4	h_3	h_1	h_2
h_3	h_1	h_4	h_2	H_x	D	H_z	H_y	h_2	h_1	h_3	h_4
h_1	h_3	h_2	h_4	H_y	H_z	D	H_x	h_3	h_4	h_2	h_1
h_2	h_4	h_1	h_3	H_z	H_y	H_x	D	h_1	h_2	h_4	h_3
h_4	h_3	h_1	h_2	h_4	h_2	h_3	h_1	D	H_x	H_y	H_z
h_2	h_1	h_3	h_4	h_3	h_1	h_4	h_2	H_x	D	H_z	H_y
h_3	h_4	h_2	h_1	h_1	h_3	h_2	h_4	H_y	H_z	D	H_x
h_1	h_2	h_4	h_3	h_2	h_4	h_1	h_3	H_z	H_y	H_x	D

Table 5.3. One of the three identical blocks of the block-diagonal Hamiltonian matrix connected with the T states

$D + H_x - H_y - H_z$	$-h_1 - h_2 + h_3 + h_4$	$-h_1 + h_2 - h_3 + h_4$
$-h_1 - h_2 + h_3 + h_4$	$D - H_x + H_y - H_z$	$+h_1 - h_2 - h_3 + h_4$
$-h_1 + h_2 - h_3 + h_4$	$+h_1 - h_2 - h_3 + h_4$	$D - H_x - H_y + H_z$

eigenvalues differ, except for accidental degeneracies. In Table 5.4 the effect of symmetry on the overlap matrix elements and the T state energies E_{T_i} is summarized. Resulting level schemes are also depicted in Fig.5.7 for the principal site symmetries. All subgroups of the tetrahedral group $\bar{4}3m$ are listed. The eigenvalues E_A and E_E are always those given in (5.3), if the above definitions for H and h are generalized to $H = \frac{1}{3}(H_x + H_y + H_z)$ and $h = \frac{1}{4}(h_1 + h_2 + h_3 + h_4)$. It may be noted that experimentally observed T state energies can be used as a sensitive probe of the site

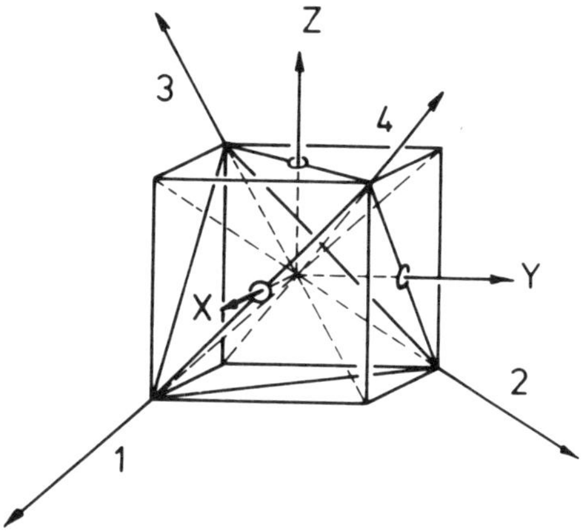

Fig.5.6. Equilibrium orientation of a tetrahedron in a rotational potential. 180° rotations around the x, y, and z axes and correspondingly 120° rotations around the 1, 2, 3, and 4 axes are the symmetry operations of the tetrahedron. Site symmetry may cause an equivalence of some of these axes

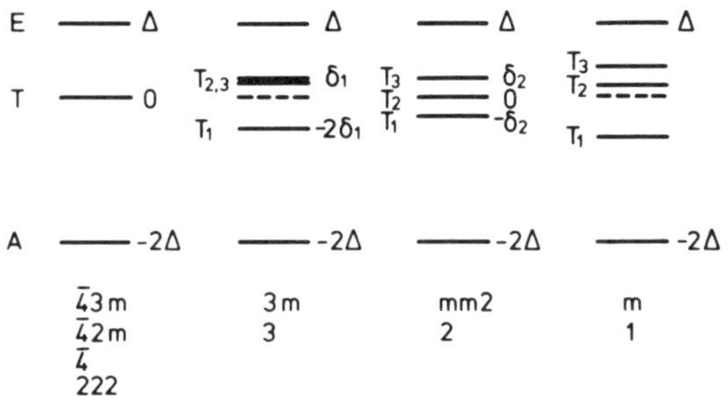

Fig.5.7. Level scheme for the rotational ground state of a tetrahedral molecule at crystal sites with different symmetry (only the 120° overlap matrix elements h_i are included). The reduction of symmetry causes a splitting of the otherwise degenerate T states. $\Delta = 4h$ denotes the main splitting, while $\delta_1 = h_2 - h_1$, and $\delta_2 = 2(h_1 - h_3)$ refer to the T state splitting for sites with three- and twofold symmetry axes, respectively. Transitions from A to E states are forbidden, as well as the T_1-T_3 transition in presence of a twofold axis. Except for very low symmetries there is always a pair of transitions with an energy ratio 2:1

symmetry; however, Table 5.4 indicates that the site symmetry in general cannot be concluded unambiguously. The results given in Table 5.4 also can be used for molecules of a symmetry lower than tetrahedral at a site with symmetry $\bar{4}3m$ (e.g., partially deuterated methane: symmetry 3m for both CHD_3 and CH_3D and symmetry mm2 for CH_2D_2).

As already noted in Sect.5.1.2, a formulation in terms of explicit wave functions means important progress. Their width (or a quantity related with it) is taken as a variational parameter. This allows the calculation of eigenvalues as a function of the rotational potential (Sect.2.6.3). Pocket states $\Phi_i = \exp(x_G \tau_i^2)$ have been used in /5.9/ and improved states in /5.14/. The latter have been qualitatively discussed before. The equilibrium orientations of the 12 pocket states are shown in Fig.5.8. For a calculation of overlap matrix elements one still has

Table 5.4. Tunneling states of tetrahedral molecules: effect of the site symmetry on the overlap matrix elements h_i and H_{xi} as well as the T state energies of a molecule with tetrahedral symmetry. Those subgroups of $\bar{4}3m$ which cause the same 120° overlap are listed together. As usually $H \ll h$ [with $H = (1/3)(H_x+H_y+H_z)$ and $h = (1/4)(h_1+h_2+h_3+h_4)$], the H_{xi} are treated in a very approximate manner in the calculation of T state levels by taking $H \equiv H_{xi}$ for all i. The effects of site symmetry and molecular symmetry can be exchanged: the relations between the overlap matrix elements also hold for tetrahedral molecules of lower symmetry in a potential of symmetry $\bar{4}3m$

subgroups of $\bar{4}3m$	120° overlap elements	180° overlap elements	T state energies E_{T_i}
$\bar{4}3m$	h h h h	H H H	
$\bar{4}2m$ $\bar{4}$	h h h h	H_x H_x H_z	E_{T_i} = D-H for all i
222	h h h h	H_x H_y H_z	
3m 3	h_1 h_2 h_2 h_2	H H H	$E_{T_i} = E_{T_2} = D-H-h_1+h_2$ $E_{T_3} = D-H+2h_1-2h_2$
mm2	h_1 h_1 h_3 h_3	H_x H_x H_z	$E_{T_i} = D-H-2(h_1-h_3)$ $E_{T_2} = D-H$
2	h_1 h_1 h_3 h_3	H_x H_y H_z	$E_{T_3} = D-H+2(h_1-h_3)$
m	h_1 h_2 h_3 h_3	H_x H_x H_z	$E_{T_{1/2}} = D-H+\frac{1}{2}(h_1+h_2-2h_3) \pm \left[2(h_1-h_2)^2+\frac{1}{4}(h_1+h_2-2h_3)^2\right]^{1/2}$ $E_{T_3} = D-H+2h_3-h_1-h_2$
1	h_1 h_2 h_3 h_4	H_x H_y H_z	from Table 5.3

to specify the potential and the kinetic energy $K = \underset{\sim}{p}^2/2\Theta$ in terms of quaternions (2.20), which reads /5.9/

$$K = -\frac{\hbar^2}{2A_1\Theta}\left\{\sum_{\alpha=1}^{d}\frac{\partial^2}{\partial\tau_\alpha^2} - A_2\sum_{\alpha=1}^{d}\tau_\alpha - \sum_{\alpha,\beta=1}^{d}\tau_\alpha\tau_\beta\frac{\partial^2}{\partial\tau_\alpha\partial\tau_\beta}\right\} . \tag{5.4}$$

For a three-dimensional rotor $A_1 = 4$, $A_2 = 3$ and $d = 4$. The generalized formulation is given because it also can be applied for one-dimensional ($A_1 = 1$, $A_2 = 1$, $d = 2$) and two-dimensional rotors ($A_1 = 1$, $A_2 = 2$, $d = 3$). In Fig.5.9 the results obtained with "improved" pocket states /5.14/ and for a potential of tetrahedral symmetry

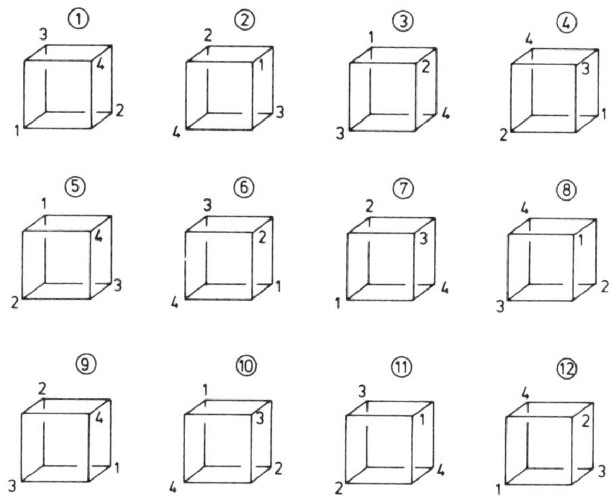

Fig.5.8

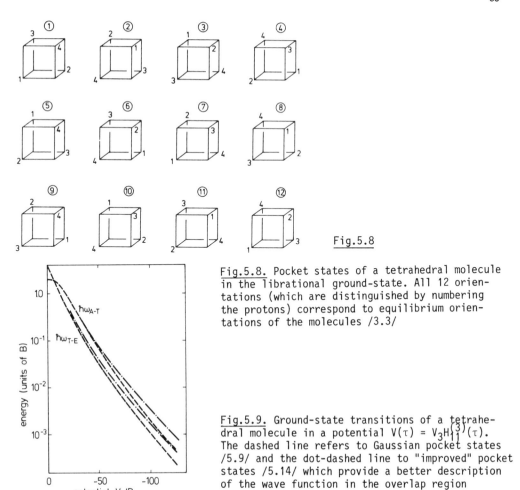

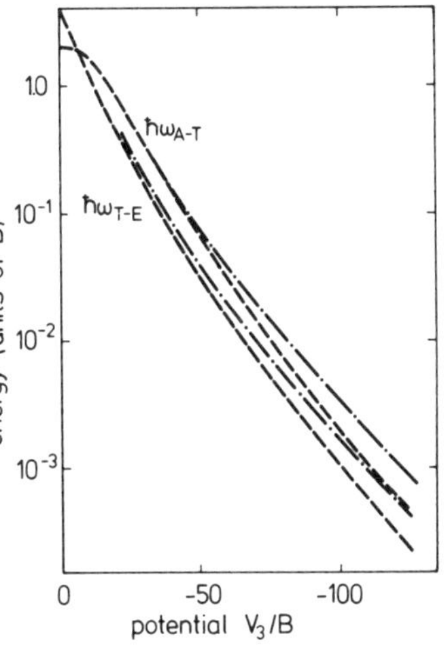

Fig.5.8. Pocket states of a tetrahedral molecule in the librational ground-state. All 12 orientations (which are distinguished by numbering the protons) correspond to equilibrium orientations of the molecules /3.3/

Fig.5.9. Ground-state transitions of a tetrahedral molecule in a potential $V(\tau) = V_3H_{11}^{(3)}(\tau)$. The dashed line refers to Gaussian pocket states /5.9/ and the dot-dashed line to "improved" pocket states /5.14/ which provide a better description of the wave function in the overlap region

$V(\tau) = V_3H_{11}^{(3)}(\tau)$ are compared with a calculation using Gaussian pocket states /5.9/. The 180° overlap essentially can be neglected, if $V_3/B \gtrsim 20$. As a general rule one may say that for pronounced orientational localization, matrix elements h_i (120° overlap) practically always dominate. Therefore, potentials which lead to an inequivalence of the 180° overlap only, in general, cause negligible T state splitting (Table 5.4). There is one example of this kind (point group $\overline{4}2m$) which becomes important in connection with the tunneling in CH_4 II (Sect.6.2.3).

As may be seen in Fig.5.9, the tunnel splitting depends nearly exponentially on the magnitude of the potential. This fact, which certainly is not restricted to tunneling of tetrahedral molecules, causes extreme sensitivity of the splitting to changes of the potential (see Sect.7.2).

We now turn to the eigenstates of a tetrahedron as calculated by expanding the wave functions into free rotor functions. A very thourough discussion of this topic was given by KING and HORNIG /5.10/ for a cubic crystal field. These authors emphasized the need to account for both molecular (symmetry operations $\hat{\bar{R}}$) and site symmetry (symmetry operations $\hat{R}$); the potential has to be invariant under the direct product group $\hat{\bar{R}}\hat{R}$ of both operations. After discussing the effect of space-fixed and molecule-fixed angular momentum operators on wave functions within the two frames, the authors constructed symmetry-adapted wave functions which have two symmetry labels. Roughly speaking, $\hat{\bar{R}}$ imposes conditions on the index m of $\mathcal{D}^{(J)}_{mm'}(\tau)$, while $\hat{R}$ acts on the index m'. $\hat{R}$ and $\hat{\bar{R}}$ represent the subgroups of proper rotations at the crystal site and in the molecule. An additional symmetry reflected in $S\,\mathcal{D}^{(J)}_{mm'}(\tau) = (-1)^{m+m'}\mathcal{D}^{(J)}_{mm'}(\tau)$ is important and originates from the direct product of the subgroups of improper rotations. Eigenvalues are calculated for a number of examples, namely potentials $V_4 H^{(4)}_{11}(\tau)$ of different magnitude. In particular the speed of convergence with the cutoff value of $J(4 \leqq J_0 \leqq 22)$ — which defines the size of the set of basis functions — is discussed. As already mentioned, convergence is relatively slow for states of lower symmetry. It is for this reason that in a series of papers by SMITH /5.21,26-28/ only the ground-state splittings with A and E symmetry are evaluated. A, for example, is a shorthand notation for $\bar{A}A$, which will be used in the following. Smith concentrated on the rotational states of NH_4^+ groups. Because of the ionicity of the crystals, the potentials usually are relatively strong in ammonium salts. This leads to convergence problems, in spite of basis sets with J_0 as large as 27. The main aim of the author is the determination of potential parameters by consistently explaining as many observations as possible, e.g., tunneling, librational states and specific heat (which is not really an independent quantity). Similar calculations have been performed by BARTHOLOME and collaborators /2.23/. They proceeded somewhat further by basing their approach on a microscopic model with electrostatic interactions.

Extensive use of the methods outlined by KING and HORNIG /5.10/ has been made by the "Kyoto group" in conjunction with solid methane and its phase transitions (/2.22/ and references therein) as well as methane as a substitutional impurity in rare-gas matrices /5.29,30/. Their calculations are based on intermolecular interactions, which represent the angle-dependent part of interactions between atoms belonging to different molecules /2.5/. This means a generalization of the famous work of JAMES and KEENAN /2.7/, who only included electrostatic multipole-multipole interactions and thus ignored the crystal field (which is present also in crystals consisting of neutral molecules). There is no electrostatic contribution to the crystal field from electrically neutral molecules. Level schemes for methane molecules and various site symmetries have been calculated and compared to experimental results /2.22,5.31/. The work by YAMAMOTO and collaborators probably represents the most advanced applications of the free rotor expansion technique.

A comparison between the pocket state approach and the expansion into free rotor functions is difficult. The speed of convergence of the latter has recently been discussed /5.14/, although mainly with respect to the excited librational states which considerably suffer from the truncation of the Hamiltonian matrix. Both approaches encounter problems for large potentials. In the free rotor function approach matrices of increasingly large dimension need to be diagonalized for reasons of convergence, while in the pocket state method the simplicity of the initial approach has to be sacrificed in order to improve the wave functions in the overlap region. Nevertheless the pocket state approach appears to be more suitable for large potentials.

5.2 Nuclear-Spin Functions

So far, only the spatial part of the wave functions has been discussed. Complete wave functions, however, must include nuclear-spin functions. In principle they also must include electronic and vibrational wave functions. At low temperatures, however, these are in the totally symmetric ground state and therefore need not to be considered here. The correct molecular wave functions are constructed from linear combinations of products of rotational and nuclear-spin functions. For protonated molecules they must be totally antisymmetric upon odd permutations of the protons (spin 1/2), whereas they must be completely symmetric upon interchange of deuterons (spin 1). An example for the antisymmetric character of the complete wave function is hydrogen. In p-H_2 the spin function χ is antisymmetric and the rotational wave function Φ is symmetric. This is reversed for o-H_2, where χ is symmetric and Φ antisymmetric. A 180° rotation of the molecule around an axis through the molecular centre-of-mass and perpendicular to the intermolecular axis is a symmetry operation of the molecule. It is equivalent to an odd permutation of two particles, namely the two protons within the molecule. This is different for XH_3 and XH_4 groups. There, odd permutations can only be performed across potential barriers which are of the order of binding energies of a molecule (several eV). Only the even permutations, which are isomorphous with the proper rotations that leave the molecule invariant, need to be considered. Consequently the complete wave functions must be totally symmetric (A symmetry) for both protonated and deuterated molecules.

This can only be achieved by combining nuclear-spin functions and rotational wave functions of the same symmetry. Transitions which change the symmetry of the rotational wave function consequently also change the symmetry of the nuclear-spin functions. As the latter cannot be caused by phonons, well-defined tunneling states with a spacing of only a few mK are observed up to about 50 K. Neutrons can flip nuclear spins and thus render the tunneling states observable. Transitions between different spin species which are not mediated by neutrons are summarized under the name spin conversion. Spin conversion is necessary to bring a mixture of several spin species into equilibrium after cooling or heating (e.g., /5.4,32/). This, however, happens on a timescale of minutes to days. In pure samples, spin conversion involves the extremely weak intramolecular dipole-dipole interaction of the nuclear magnetic moments /5.33/. It is known that conversion is speeded up considerably by the presence of paramagnetic impurities /5.32/.

For three protons forming an equilateral triangle (Fig.2.2) as in CH_3 groups or NH_3 (point group 3m), there are 2^3 ways of arranging the nuclear spins in $|\mu_1\mu_2\mu_3\rangle$. Here the μ_i denote the z component of the nuclear spin of the ith proton. The resulting eight spin functions can be decomposed into four totally symmetric states χ_A with total nuclear spin I = 3/2 of the molecule and two pairs of doubly degenerate functions χ_E with total spin 1/2 (Fig.5.10). The eight spin functions are listed in Table 5.5. It may be noted that the total spin I of the molecule unambiguously labels the symmetry in the present case.

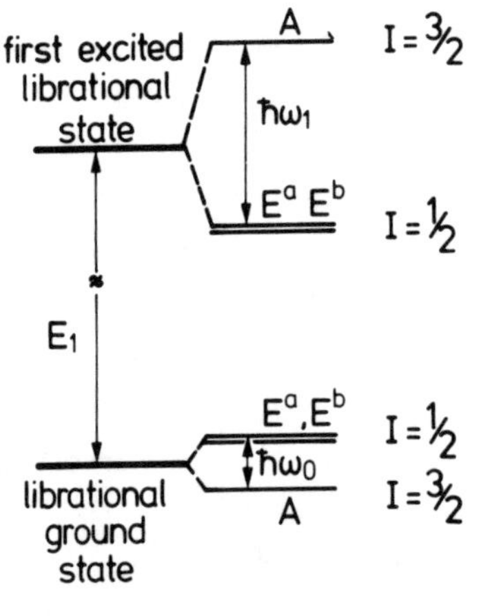

Fig.5.10. Low-energy rotational states of a XH_3 group, labeled by their symmetry. In general $E_1 \gg \hbar\omega_1 \gg \hbar\omega_0$; note that the A state has lower energy than the E state in the ground-state levels; this is reversed in the first excited librational state

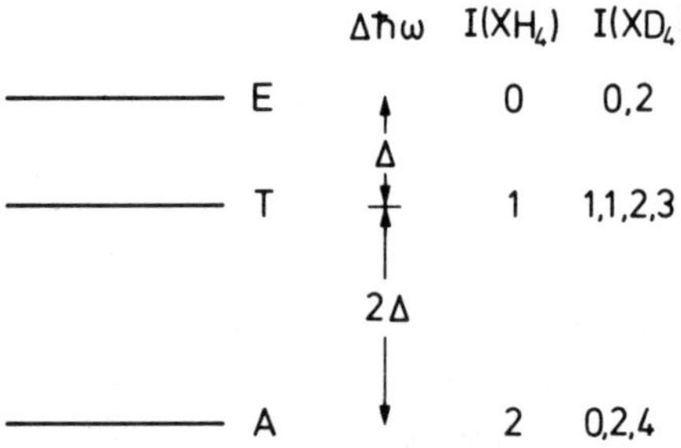

Fig.5.11. Ground-state multiplet for tetrahedral XH_4 and XD_4 molecules in a potential of tetrahedral symmetry. For lower symmetry a splitting of the degenerate T states results (see Table 5.4 and Fig.6.16). The transitions observed by spin-incoherent neutron scattering are marked by arrows; I denotes the total nuclear spin of the respective molecule

Table 5.5. Spin functions $\chi(\Gamma_s, m_I)$ of CH_3 groups and NH_3 molecules. I denotes the total nuclear spin of a molecule, m_I its z component, $\alpha = 1/2$ and $\beta = -1/2$ the z component of the proton spin, Γ_s labels the symmetry of the wave functions and $\varepsilon = -1/2 + i\sqrt{3}/2$

	m_I	$\chi(\Gamma_s, m_I)$
type A	3/2	$\alpha\alpha\alpha$
	1/2	$(1/\sqrt{3})(\alpha\alpha\beta + \alpha\beta\alpha + \beta\alpha\alpha)$
I = 3/2	-1/2	$(1/\sqrt{3})(\beta\beta\alpha + \beta\alpha\beta + \alpha\beta\beta)$
	-3/2	$\beta\beta\beta$
type E	1/2	$(1/\sqrt{3})(\alpha\alpha\beta + \varepsilon\cdot\beta\alpha\alpha + \varepsilon^*\cdot\alpha\beta\alpha)$
	-1/2	$(1/\sqrt{3})(\beta\beta\alpha + \varepsilon\cdot\alpha\beta\beta + \varepsilon^*\cdot\beta\alpha\beta)$
I = 1/2	1/2	$(1/\sqrt{3})(\alpha\alpha\beta + \varepsilon^*\cdot\beta\alpha\alpha + \varepsilon\cdot\alpha\beta\alpha)$
	-1/2	$(1/\sqrt{3})(\beta\beta\alpha + \varepsilon^*\cdot\alpha\beta\beta + \varepsilon\cdot\beta\alpha\beta)$

Table 5.6. Spin functions $\chi(\Gamma_s, m_I)$ of tetrahedral molecules XH_4. I denotes the total nuclear spin of the molecule, m_I its z component, $\alpha = 1/2$ and $\beta = -1/2$ the z component of the proton spins, Γ_s labels the symmetry of the wave functions and $\varepsilon = -1/2 + i\sqrt{3}/2$

	m_I	$\chi(\Gamma_s, m_I)$
type A I = 2	2	$\alpha\alpha\alpha\alpha$
	1	$(1/2)(\alpha\alpha\alpha\beta + \alpha\alpha\beta\alpha + \alpha\beta\alpha\alpha + \beta\alpha\alpha\alpha)$
	0	$(1/\sqrt{6})(\alpha\alpha\beta\beta + \alpha\beta\alpha\beta + \alpha\beta\beta\alpha + \beta\alpha\alpha\beta + \beta\alpha\beta\alpha + \beta\beta\alpha\alpha)$
	-1	$(1/2)(\beta\beta\beta\alpha + \beta\beta\alpha\beta + \beta\alpha\beta\beta + \alpha\beta\beta\beta)$
	-2	$\beta\beta\beta$
type T I = 0	1	$(1/2)(\alpha\beta\alpha\alpha + \alpha\alpha\beta\alpha - \beta\alpha\alpha\alpha - \alpha\alpha\alpha\beta)$
T_x	0	$(1/2)(\alpha\beta\beta\alpha - \beta\alpha\alpha\beta)$
	-1	$(1/2)(-\beta\alpha\beta\beta - \beta\beta\alpha\beta + \beta\beta\beta\alpha + \alpha\beta\beta\beta)$
	1	$(1/2)(\beta\alpha\alpha\alpha + \alpha\alpha\beta\alpha - \alpha\alpha\alpha\beta - \alpha\beta\alpha\alpha)$
T_y	0	$(1/\sqrt{2})(\beta\alpha\beta\alpha - \alpha\beta\alpha\beta)$
	-1	$(1/2)(\alpha\beta\beta\beta + \beta\beta\alpha\beta - \beta\beta\beta\alpha - \beta\alpha\beta\beta)$
	1	$(1/2)(\beta\alpha\alpha\alpha + \alpha\beta\alpha\alpha - \alpha\alpha\alpha\beta - \alpha\alpha\beta\alpha)$
T_z	0	$(1/\sqrt{2})(\beta\beta\alpha\alpha - \alpha\alpha\beta\beta)$
	-1	$(1/2)(\alpha\beta\beta\beta + \alpha\beta\alpha\alpha - \alpha\alpha\alpha\beta - \alpha\alpha\beta\alpha)$
type E I = 0	0	$(1/\sqrt{6})(\alpha\alpha\beta\beta + \beta\beta\alpha\alpha + \varepsilon[\alpha\beta\alpha\beta + \beta\alpha\beta\alpha] + \varepsilon^*[\alpha\beta\beta\alpha + \beta\alpha\alpha\beta])$
	0	$(1/\sqrt{6})(\alpha\alpha\beta\beta + \beta\beta\alpha\alpha + \varepsilon^*[\alpha\beta\alpha\beta + \beta\alpha\beta\alpha] + \varepsilon[\alpha\beta\beta\alpha + \beta\alpha\alpha\beta])$

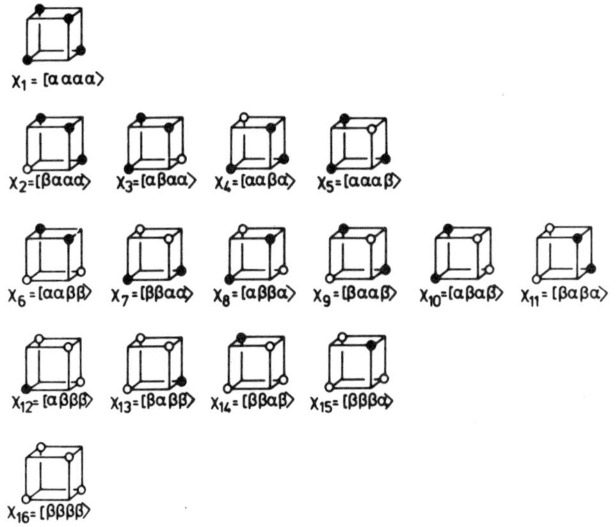

Fig.5.12. Wave functions $[\mu_1\mu_2\mu_3\mu_4>$ which define the spin states of the protons at positions i in the crystal, and not the spin of the ith proton. Open circles denote spin up, full circles spin down /3.3/

For tetrahedral molecules XH_4 there are 16 spin functions and they may be decomposed as $\chi = 5\ \chi_A + \chi_E + 3\ \chi_T$. The spin functions /5.34/ are listed in Table 5.6. Again the total nuclear spin is related to the symmetry in a one-to-one relation (Fig.5.11). This is different for the fully deuterated species (e.g., CD_4 with 81 spin functions). Nuclear-spin functions for XD_3 and XD_4 may be found in a paper by HÖLLER and PRESS /3.4/. Their construction requires the application of projection operators (in order to construct properly symmetrized function), and use of the fact that these functions are eigenfunctions of both $\mathring{i}_z$ and $\mathring{i}^2$. This is described in more detail in the appendix.

Correctly symmetrized complete wave functions for orientationally ordered tetrahedral XH_4 have been constructed by combining the 12 rotational wave functions and the 16 nuclear-spin functions into 192 products $|\Phi_i>|\mu_1\mu_2\mu_3\mu_4>$. The method is outlined in /3.3/ which also gives a table of these symmetrized states. It should be noted that these states are not eigenstates of the molecule in a given potential. Eigenstates are obtained by diagonalization of the Hamiltonian matrix $\mathscr{H}_T$. HÖLLER /3.3/ gives a simple interpretation of the 16 symmetrized wave functions in terms of new wave functions $[\mu_1\mu_2\mu_3\mu_4>$. Here μ_i does not denote the spin of *particle i*, but that of the nucleus which is in *position i*. $[\beta\alpha\alpha\alpha>$, for example, means that the nuclear spin with z component $\beta \equiv -1/2$ is at site 1. All 16 functions $[\mu_1\mu_2\mu_3\mu_4>$ are shown in Fig.5.12. They fulfill the symmetry requirements upon particle exchange and are particularly useful for the calculation of transition matrix elements (Sects.3.1, 5.3). An equivalent way consists in first constructing symmetrized spatial wave functions Φ_Γ and symmetrized spin functions χ_Γ. Then functions of the same symmetry are combined to totally symmetric wave functions $\psi = \Phi_\Gamma \cdot \chi_\Gamma$.

The combination of spatial and spin functions for free rotor states of XH_4 molecules has been demonstrated by HAMA and MIYAGI /3.6/.

5.3 Transition Matrix Elements for Neutron Scattering

With knowledge of the complete molecular wave functions, there is no further conceptual difficulty in evaluating the transition matrix elements which appear in (3.10). Three recent publications deal with this problem /3.3,6.5,35/ and all refer to tetrahedral (spherical top) molecules with their explicit formulation. An application to CH_3 group rotation is straightforward, and a summary of the results is given in the appendix.

HAMA and MIYAGI /3.6/ studied the limit of free rotor states, HÜLLER /3.3/ approximates the rotational wave functions by pocket states of zero librational amplitude (which recently has been generalized to finite librational amplitudes /3.4/). OZAKI et al. /5.35/ dealt with the intermediate regime and mainly base on the formalism developed by HAMA and MIYAGI /3.6/. The final results of /3.6,5.35/ look rather complex and necessitate the introduction of numerous group-theoretical definitions. Therefore only some steps in their derivation will be outlined.

As it appears, the main difficulty in the evaluation of matrix elements is the proper inclusion of symmetrized wave functions. The results should reflect correlations of the protons within a molecule which give rise to interference effects /3.3,6/. HAMA and MIYAGI noted that previous publications /3.5,10,5.36-38/, in which the transition matrix elements are calculated for free molecules (in the gas phase), either omitted these correlations or did not include them correctly.

The common aim of all three publications is the calculation of the transition matrix elements $A^n_{\mu'a',\mu a}$ which appear in (3.10)

$$A^n_{\mu'a',\mu a} = \sum_{\gamma=1}^{4} \langle\mu'\psi_{na'}|W^{n\gamma}|\mu\psi_{na}\rangle \tag{5.5a}$$

with

$$W^{n\gamma} = (A^{n\gamma}-\bar{A})\exp(i\underline{Q}\cdot\underline{r}_{n\gamma}) \quad . \tag{5.5b}$$

The meaning of the quantities in (5.5) has been explained in Sect.3.1. In the following the index n will be dropped and reference is to just one molecule.

Let us first consider the approaches based on free rotor functions /3.6/ or an expansion into these functions /5.35/. The emphasis will be on the second paper. In order to calculate $A_{\mu'a',\mu a}$ it is necessary to introduce three coordinate systems: 1) a space-fixed frame with coordinate $\underline{X}$, 2) a crystal frame $\underline{X}_C$ with orientation ω_2^E with respect to $\underline{X}$, and 3) a molecular frame $\underline{X}_M$ with orientation ω_1^E with respect to $\underline{X}_C$. For a potential $V_{st} = 0$ (which corresponds to the approach of HAMA and MIYAGI /3.6/) there is no preferred orientation of the molecule in the crystal and obviously there is no need of distinguishing between $\underline{X}$ and $\underline{X}_C$. Furthermore the polar coordinates of the vectors $\underline{Q}$ and $\underline{r}_\gamma$ are introduced. They are Ω_Q and Ω_γ in the space-fixed frame and Ω_Q^M and Ω_γ^M in the molecular frame.

As the wave functions $|\mu\Phi_a>$ need to be expressed in symmetry-adapted functions /3.3,6,5.35/, it is convenient also to express the neutron scattering operator

$$W = \frac{2a_{inc}}{\sqrt{I(I+1)}} \sum_{\gamma=1}^{4} \mathring{s}\cdot\mathring{i}_\gamma \exp(i\underline{Q}\cdot\underline{r}_\gamma) \tag{5.6}$$

in terms of symmetry-adapted functions. The tetrahedral symmetry of the molecules enters in the following way:

$\tilde{W} = \sum_{\gamma=1}^{4} \mathring{s}\mathring{i}_\gamma G_\gamma$ with $G_\gamma = \exp(i\underline{Q}\cdot\underline{r}_\gamma)$ may be decomposed into /5.39/

$$\tilde{W} = \tilde{W}_A+\tilde{W}_T = \tilde{W}_A+\tilde{W}_{T_x}+\tilde{W}_{T_y}+\tilde{W}_{T_z} \tag{5.7}$$

with

$$\tilde{W}_A = \frac{1}{4}\,\mathring{s}(\mathring{i}_1+\mathring{i}_2+\mathring{i}_3+\mathring{i}_4)\sum_{\gamma=1}^{4} G_\gamma = \frac{1}{4}\,\mathring{s}\cdot\mathring{i}_{tot}\sum_{\gamma=1}^{4} G_\gamma \tag{5.8a}$$

$$\tilde{W}_{T_x} = \frac{1}{4}\,\mathring{s}(\mathring{i}_1-\mathring{i}_2-\mathring{i}_3+\mathring{i}_4)(G_1-G_2-G_3+G_4) \tag{5.8b}$$

$$\tilde{W}_{T_y} = \frac{1}{4}\,\mathring{s}(\mathring{i}_1-\mathring{i}_2+\mathring{i}_3-\mathring{i}_4)(G_1-G_2+G_3-G_4) \tag{5.8c}$$

$$\tilde{W}_{T_z} = \frac{1}{4}\,\mathring{s}(\mathring{i}_1+\mathring{i}_2-\mathring{i}_3-\mathring{i}_4)(G_1+G_2-G_3-G_4) \quad . \tag{5.8d}$$

$\tilde{W}_A$ and $\tilde{W}_T$ are irreducible forms within the group of even permutations of four elements (point group T = 23). It is noted that W does not contain a component $\tilde{W}_E$ of the operator with E symmetry. The above expressions are obtained by use of projection operators, in the same way as in conjunction with spin functions (see appendix).

A similar formulation has been given by SARMA for the neutron scattering in ortho- and para-hydrogen /5.40/. HAMA and MIYAGI proceeded in a somewhat different way /3.6/. They expressed the spin and spatial parts of (5.6) separately in terms of symmetry-adapted operators.

What is important, though, is the absence of a component of the neutron scattering operator W with E symmetry. As can easily be found by inspection of the multiplication table of the point group T, matrix elements $\langle\chi_A\Phi_A|W|\Phi_E\chi_E\rangle$ require a component of the neutron scattering operator with E symmetry. As there is no such component, matrix elements $\langle\chi_A\Phi_A|W|\Phi_E\chi_E\rangle$ vanish. We will return to this aspect when describing the results of neutron scattering experiments (Chap.6).

The Euler angles ω_1^E and ω_2^E come into play when $\exp(i\underline{Q}\cdot\underline{r}_\gamma)$ in (5.6) is expanded within the space-fixed frame $\underline{X}$

$$\exp(i\underline{Q}\cdot\underline{r}_\gamma) = 4\pi \sum_{\ell=0}^{\infty} i^\ell j_\ell(Q\rho) \sum_{m=-\ell}^{\ell} Y^*_{\ell m}(\Omega_Q)Y_{\ell m}(\Omega_\gamma) \quad . \tag{5.9}$$

j_ℓ denotes spherical Bessel functions. Then the spherical harmonic $Y_{\ell m}(\Omega_\gamma)$ is transformed to the molecular frame

$$Y_{\ell m}(\Omega_\gamma) = \sum_{k=-\ell}^{\ell} \sum_{m'=-\ell}^{\ell} Y_{\ell k}(\Omega_\gamma^M)\mathcal{D}^{(\ell)}_{km'}(\omega_1^E)\mathcal{D}^{(\ell)}_{m'm}(\omega_2^E) \tag{5.10}$$

In order to simplify the expression for $A_{\mu'a',\mu a}$ (5.5) a powder average is performed /5.35/, which means an integration over the Euler angles ω_2^E. The result may be written in terms of an intermediate scattering function $I(Q,t)$ which formally looks identical to (4.5)

$$I(Q,t) = \sum_{\ell=0}^{\infty} (2\ell+1)j_\ell(Q\rho)^2 F_\ell(t) \tag{5.11}$$

with

$$F_\ell(t) = \sum_{i,f} p_i(T)\,\frac{64}{3}\,G_\ell^{i,f} \quad . \tag{5.12}$$

The summation is over all initial and final states. $G_\ell^{i,f}$ is a transition matrix element which only contains the angular part (angle ω_1^E) of $A_{\mu'a',\mu a}$. Proper account has to be taken of the degeneracies of the states. For the 0-1 rotational transition of a free molecule, e.g., the intensity is simply proportional to $j_1^2(Q\rho)$. Concerning explicit expressions of the matrix elements $G_\ell^{i,f}$ and concerning applications the reader is referred to /3.6,5.35/. HAMA and MIYAGI were mainly interested in correlation effects /3.6/, while OZAKI et al. /5.35/ applied their calculation to the neutron spectra of CH_4 II /5.4,41/.

Here one should note that the correlation effects are due to correlations of nuclear spins *within* a molecule. An apparent effect concerns the scattering intensity at momentum transfer Q = 0. For a sample consisting only of A species molecules (temperature << tunnel splitting), this intensity is 8 σ_{inc}. This means a doubling in comparison to the scattering intensity from a high-temperature mixture of spin species (all states — 5A, 3*3T, 2E — equally populated) or in the classical limit. Both yield 4 σ_{inc}. On the other hand, the scattering from different molecules does not give rise to additional correlation effects (even if completely converted to the A species), as long as the z components of the total nuclear-spins of different molecules are statistically independent.

Nuclear-spin ordering appears possible, however, either statistically at low temperature (order of μK) and/or high magnetic fields or by means of dynamical polarization.

The intensities of observed rotational transitions yield information which may complement that from peak positions. This may be seen in analogy to the measurement of phonon modes. Peak positions yield information on the phonon frequencies, while $\underline{Q}$-dependent intensities allow one to learn about the symmetry and magnitude of the displacement vectors (mode eigenvectors). Similarly the inclusion of intensities may help to assign the symmetry of rotational states /5.15/. As calculations of matrix elements only recently have become available, little has yet been done along these lines.

As with the calculation of rotational energy eigenvalues there is an alternative approach to the calculation of matrix elements $A_{\mu'a',\mu a}$ within the pocket state formalism /3.3/. HÜLLER first calculated matrix elements $B_{\mu'b',\mu b}$ based on (unsymmetrized) pocket states (see preceding section) and then transformed to eigenstates by means of a unitary matrix. He used δ functions, that is, pocket states of zero width. This approximation is performed in view of an application of the results to ammonium salts with small librational amplitude of the NH_4^+ group and the present restriction of high-resolution neutron spectroscopy to $Q \lesssim 1.8$ Å^{-1}. The approach, therefore, breaks down for large Q and large librational amplitudes. This limitation is not serious, however. A calculation based on the pocket state formalism but including the finite width of these states has recently been performed /3.4/. A simple approach as in /3.4/ allows the study of scattering as a function of the librational amplitude which cannot easily be extracted from /5.35/. On the other hand, the calculation is applicable for moderately large momentum transfers and librational amplitudes only.

The calculation of transition matrix elements /3.3,4/ on the basis of pocket states is outlined in the appendix. It is distinguished between a) spin-flip scattering (operators $s_+i_{\gamma-}$ and $s_-i_{\gamma+}$) and b) non-spin-flip scattering (operator $s_zi_{\gamma z}$). Inelastic scattering is caused by the part of the neutron scattering operator with T symmetry. For degenerate T states there is also a contribution to the elastic scattering, originating from the operator W_T.

Only the part of the elastic scattering which results from the totally symmetric neutron scattering operator W_A also corresponds to the elastic intensity in the classical high-temperature limit. This relation probably has direct consequences for the understanding of the temperature dependence (Sect.7.1).

One should note that the matrix elements $B_{\mu'b',\mu b}$ are derived for trigonal site symmetry $C_3(\equiv 3)$. This is related to the fact that the calculations in /3.3/ are performed in view of the tunneling in NH_4ClO_4. The accidental degeneracy of T_2 and T_3 states in NH_4ClO_4 means that the true site symmetry (m) can be replaced by an effective trigonal symmetry (see also Table 5.4). The tables in /3.4/ and in the appendix mainly refer to cubic symmetry.

The results initially are given for single crystals, whereby the orientation of the molecules is introduced by the set of vectors $\underline{r}_\gamma$ (denoting the positions of the atoms of the corners of a tetrahedron). Interference effects in the scattering caused by correlations of the proton are predicted, as by HAMA and MIYAGI /3.6/. As a test neutron scattering experiments with single crystals are suggested /3.3/, preferably for systems with no T state degeneracy. In contrast to this /3.5,6/ refer to the hypothetical situation of a methane gas at very low temperatures.

In agreement with the derivation given before, it is found that A-E transitions are forbidden. For the purpose of a direct comparison with the scattering from powder samples, powder averages of the calculated intensities have to be performed. Results for tetrahedral symmetry—but also for reduced site symmetry—are given in the appendix. Rather good agreement between theory and experiment is obtained (see Sect. 6.2.4).

Also /3.3/ has served as a basis for the calculation of the total neutron scattering cross section σ_{tot} /5.42/. σ_{tot} represents an integration over all allowed transitions which enter the double differential cross section. Usually this integral information is obtained by transmission experiments with long-wavelengths neutrons, hence Q is small and the approximation of negligible librational amplitude appears permissible. As σ_{tot} depends strongly on the population of the librational ground-state levels, it is well-suited for measuring nuclear-spin conversion with long conversion times.

6. Rotational Excitations at Low Temperatures
II. Examples

After having discussed the theoretical background, we now turn to examples for the observation of rotational excitations at low temperatures. Only very few examples close to the limit of free rotation exist, whereas there is a fairly large and steadily increasing number of experiments in the limit of rotational tunneling. The chapter mainly is based on experiments with inelastic spin-dependent neutron scattering. Other experimental techniques may yield complementary information and also will be mentioned in a less detailed fashion. Examples with a ground-state splitting not accessible to neutron scattering will not be included.

6.1 Free Rotation

As mentioned before, there are only few molecular solids close to the limit of free rotation. This is due to the intermolecular interaction in a crystal which is practically always much bigger than the rotational constant B. The only exception with $B \gg V$ is solid hydrogen: there B = 85.25 K = 7.35 meV compares with an anisotropic interaction (electrostatic quadrupole-quadrupole interaction) which gives rise to an ordering transition at about 3 K for o-H_2. In all other examples $B < V$. If a large spacing of the low-lying states is observed nevertheless, it is caused by a combination of 1) high symmetry and 2) moderate strength of the potential.

This can be illustrated with potentials of the form $V(\phi) = \frac{1}{2}V_n \cos n\phi$, for which the calculation of eigenvalues is particularly simple. For the uniaxial rotation of a dumbbell, low symmetry means $V_2 \neq 0$. High symmetry, on the other hand, means $V_2 = 0$ and maybe $V_4 \neq 0$ or $V_6 \neq 0$. Eigenvalues with the same splitting are obtained for different V_n, if the magnitude of the potential V_n is increased by a factor $(n/2)^2$ with respect to V_2. Consequently for the same magnitude $V_n = V_2$ $(n > 2)$ the perturbation of the free states is much weaker (for a comparison of the eigenvalues as a function of V_3 and V_6, see Fig.5.4). Within the pocket state picture, more symmetry means more closely spaced potential minima and therefore greater overlap. All presently known examples for almost free rotation will be given below.

6.1.1 Solid Hydrogen

Though solid hydrogen is the prime example of free rotation in the solid, we shall treat it only briefly and mainly refer to the literature /2.11,3.12,5.1,2,6.1,2/. Experiments have been performed both with p-H_2 /5.1,6.1,2/ and with crystals highly enriched with o-H_2 /3.12,5.2/. p-H_2 molecules at low temperatures are in the rotational ground state and the hexagonal crystalline field is very weak. The 0-1 transition is observed at E_{0-1} = 14.6 meV /5.1,6.1/ which compares to a value of 2B = 14.7 meV for the free molecule. Only with the anisotropic ortho species (I = 1, J = 1) and for o-H_2 concentrations $x > 55\%$ is a phase transition from the disordered hcp phase to an orientationally ordered cubic phase (4 sublattices, space group Pa3 /6.3,4/) observed. Reduced E_{1-0} values of about 14 meV /3.12/ mainly are due to the additional molecular field in the ordered phase. In cubic o-H_2 the J = 1 state splits into a ground state described by a wave function $Y_{10}(\Omega)$ (with its axis along body diagonals) and two degenerate excited states $Y_{1\pm1}(\Omega)$, which propagate like spin waves (librons). Unfortunately, the libron structure factor for inelastic neutron scattering is very unfavorable /6.5/. Librons, however, have been observed directly with optical techniques (references in /3.12,6.5/). Another very interesting phenomenon concerns the observation of spectra of o-H_2 pairs in p-H_2 matrices /6.6/.

6.1.2 Methane (CH_4)

At low temperatures, CH_4 is in a partially ordered phase II (with respect to the molecular orientations). The structure has been predicted by JAMES and KEENAN /2.7/ on the basis of electrostatic octopole-octopole interactions (EOO). It has been verified by a neutron diffraction experiment /2.12/ for CD_4 and afterwards a more indirect confirmation by inelastic neutron scattering experiments has been given for CH_4, too /5.5/. The 8-sublattice structure is shown in Fig.6.1. It consists of 6 ordered sites with point symmetry $\bar{4}2m$ and 2 disordered sites with point symmetry 432. The disorder in the latter case is due to a cancellation of the EOO interaction; this gives rise to a fourfold axis at the crystal site — which the tetrahedral molecules does not have.

At all sites a cubic crystalline field $V_c(\omega^E)$ (~ 150 K) is present (see also Sect.6.1.3), whereas the molecular field, stemming from the EOO interaction, only acts on the ordered molecules. The excitations of the ordered molecules will be

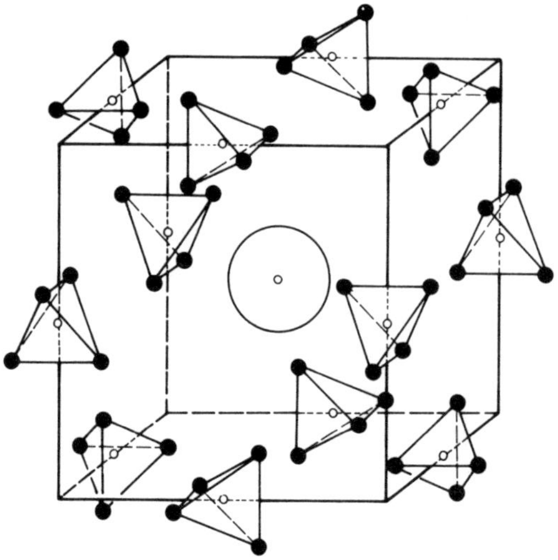

Fig.6.1. Structure of phase II of methane (space group Fm3c). There are 6 sublattices with orientationally ordered molecules (site symmetry $\bar{4}2m$) and 2 sublattices with orientationally disordered molecules (site symmetry 432). One of the disordered molecules surrounded by a cage of ordered neighbors is shown /2.12/

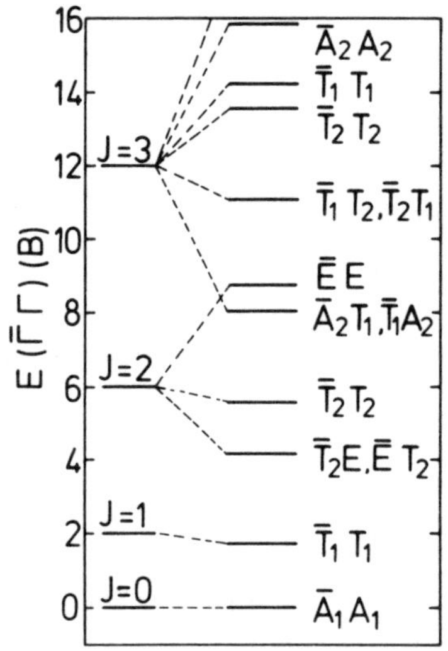

Fig.6.2

Fig.6.3

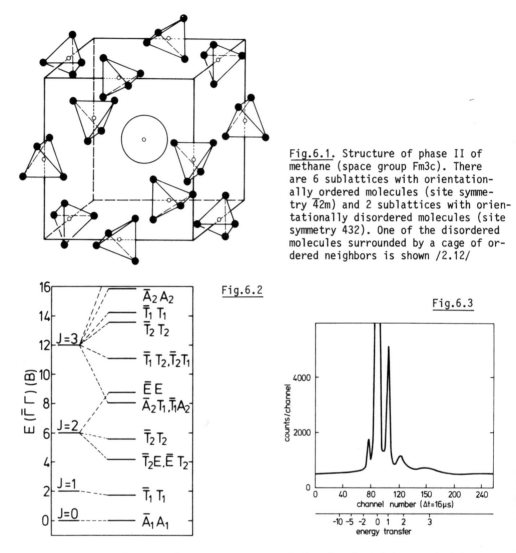

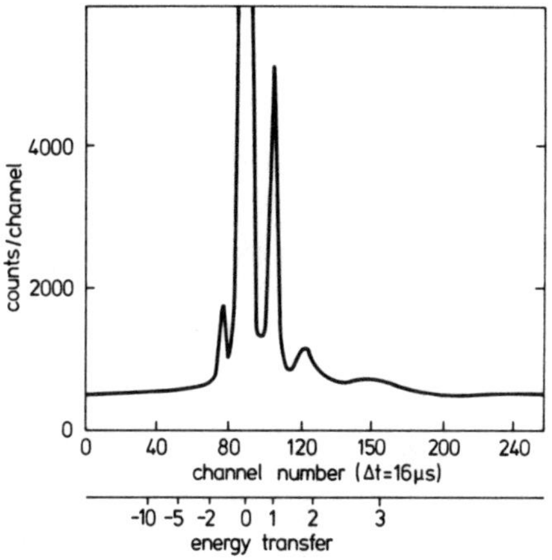

Fig.6.2. Rotational energy levels of a tetrahedral molecules without and with a crystalline field of octahedral symmetry (potential parameters as for disordered molecules in CH_4 II)

Fig.6.3. Almost free rotation in solid CH_4 II (disordered molecules /5.4/). The line is a smoothened representation of the experimental data

described in Sect.6.2.3. Therefore the appearance of almost free rotors in CH_4 II mainly is due to the symmetry at the site of disordered molecules.

The rotational states of a tetrahedral rotor (Sect.5.1.4) with potential parameters determined for CH_4 are shown in Fig.6.2. Measurements with a resolution ΔE = 0.2 meV have been performed by KAPULLA and GLÄSER /5.3,4/ and also by PRESS and KOLLMAR /5.5/. In /5.4/ levels at 1.06 meV, 1.8 meV and 2.8 meV (Fig.6.3) are found and have been interpreted as 0-1, 1-2 and 0-2 transitions, respectively (rotational constant B = 0.65 meV). As may be seen from Fig.6.2, there is some ambiguity concerning the assignment of a symmetry label to the observed (perturbed) J = 2 states. This is very troublesome in quantitative analyses /5.15/, which have to rely on a correct assignment.

States $\psi_{E\bar{E}}$ cannot be reached from the ground state $\psi_{A\bar{A}}$, for reasons of nuclear-spin conservation (see Sect.5.3). Energy levels determined by an expansion into free rotor functions /5.31/ agree rather well with the experimental results. A crystal field of the form $V_c(\omega^E) = V_4 U_{11}^{(4)}(\omega^E) + V_6 U_{11}^{(6)}(\omega^E)$ has been used in these calculations.

Whereas information on the level scheme of a rotor is contained in the positions of the inelastic peaks, the detailed shape of the wave functions determines the $\underline{Q}$ dependent intensity of the observed peaks. KAPULLA and GLÄSER have compared the Q dependence of the experimentally observed scattering intensity I_{0-1} of the 0-1 transition /5.4/ (powder samples) with results of a calculation based on free rotor functions /6.7/. Qualitatively the agreement is rather good. There is a slight displacement of the maximum of the measured intensity towards smaller Q, compared with the theoretical curve. This effect probably is spurious, however, because the admixture of higher order free rotor functions tends to shift the maximum in the opposite direction. Very recently, transition matrix elements have been calculated by OZAKI et al. /5.35/ (see also Sect.5.3), who use perturbed free rotor functions. The agreement with the previous results /6.7/ is surprisingly good, which indicates that the admixture of higher states has relatively little effect on the Q dependence of the 0-1 intensity. The treatment in /5.35/ closely follows the formalism developed by HAMA and MIYAGI /3.6/.

KAPULLA and GLÄSER /5.4/ performed measurements with samples containing 1% oxygen impurities (in order to speed up spin conversion) and with pure samples. In the latter case thermal equilibrium is reached very slowly (order of several hours) as can be judged from the time dependence of I_{0-1}/I_{1-0} or — what is equivalent — the time dependence of an effective temperature introduced via detailed balance (Sect.3.2): Slow conversion of the disordered molecules has not been confirmed by NMR /5.32/. An attempt has been made to explain spin conversion in CH_4 by a model based on intramolecular dipole-dipole interaction in conjunction with the intermolecular EOO interaction /5.33/. The magnetic dipole-dipole interaction mixes the spin states, while the EOO interaction (modulated by phonons) is responsible for rotational transitions. The acoustic phonon density of states at the rotational energy enters. It is proportional to ω^2 and, accordingly, conversion is fast at disordered sites and slow on ordered ones. Qualitatively this prediction is in agreement with NMR /5.32/ and not with the results obtained by inelastic neutron scattering /5.4/.

At pressures $p > 300$ bar, CH_4 transforms into its phase III /6.8/, which presumably is fully ordered with respect to the molecular orientations /2.12,6.9/. This is supported by a recent neutron scattering experiment at $p = 800$ bar /6.10/. No inelastic peak is observed in the region $0.2 \leq E \leq 2$ meV. Therefore almost free rotations of CH_4 molecules in phase III can be ruled out. The formerly disordered molecules probably display a tunnel splitting at energies $E < 0.2$ meV.

Another interesting feature concerns the temperature dependence of the rotational states. Obviously the line spectra must merge into one broad quasielastic component (connected with rotational diffusion) above the ordering transition at 20.4 K. This already has been observed in /5.4/. More detailed measurements /5.5/ show that the energy of the 0-1 transition shifts only little, while the linewidth of the observed peaks rises rather steeply with temperature (Fig.6.4). Lacking a better description of the scattering function, the positions and widths have been obtained by fitting Lorentzians to the data (see Sect.6.2.3a).

6.1.3 CH_4 in Rare-Gas Matrices

If it is true that the disordered molecules in CH_4 II only experience a crystalline field and no molecular field (just referring to V_{st}), rotational levels of a very similar spacing should be observed for CH_4 molecules as substitutional impurities in rare-gas matrices.

The angle-independent parts of the interaction between two methane molecules and between CH_4 and a rare-gas atom (Ar, Kr, Xe) are not very different. Therefore the two partners mix rather well, and local lattice relaxations which locally perturb the free lattice are comparatively small. Low concentrations of CH_4 would be desirable to avoid a direct interaction between methane molecules, but for intensity reasons CH_4 concentrations of ~1% are required. Results with argon, krypton, and xenon matrices (all fcc) are shown in Fig.6.5 /5.15/. These spectra indeed display features very similar to those observed in CH_4 II. They have been complemented by other measurements with relaxed resolution in order to include rotational states of higher energies. The observed energies /5.15/ for the perturbed 0-1 transition range from $E_{0-1}(Ar) = 0.89$ meV, $E_{0-1}(Kr) = 0.98$ meV, $E_{0-1}(CH_4) = 1.07_5$ meV (average between E_{0-1} from /5.4,5/) to $E_{0-1}(Xe) = 1.13_5$ meV. They rise monotonically from argon to xenon, as do the lattice constants of the matrices (Fig.6.6). This seems to prove that the static potential V_{st} in CH_4 (sites of disordered molecules) can be identified with the crystalline field $V_c(\omega^E)$. Free rotation ($E_{0-1} = 1.30$ meV) is approached more closely, the larger the "cage" in which a CH_4 molecule rotates.

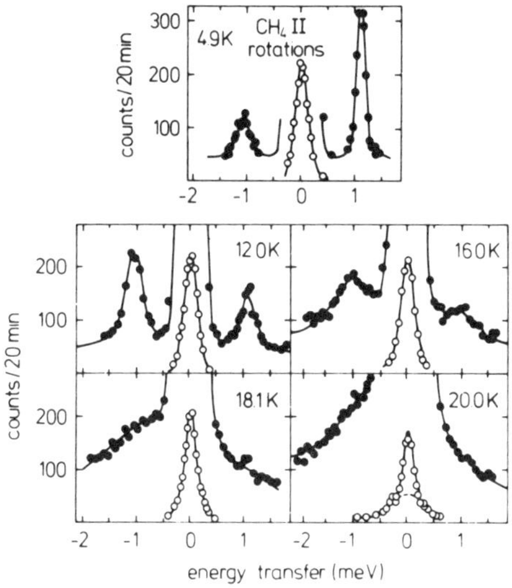

Fig.6.4. Temperature dependence of the low-energy rotational excitations of the disordered molecules in CH_4 II; the neutron scattering data have been measured with an energy resolution of about 0.2 meV /5.5/

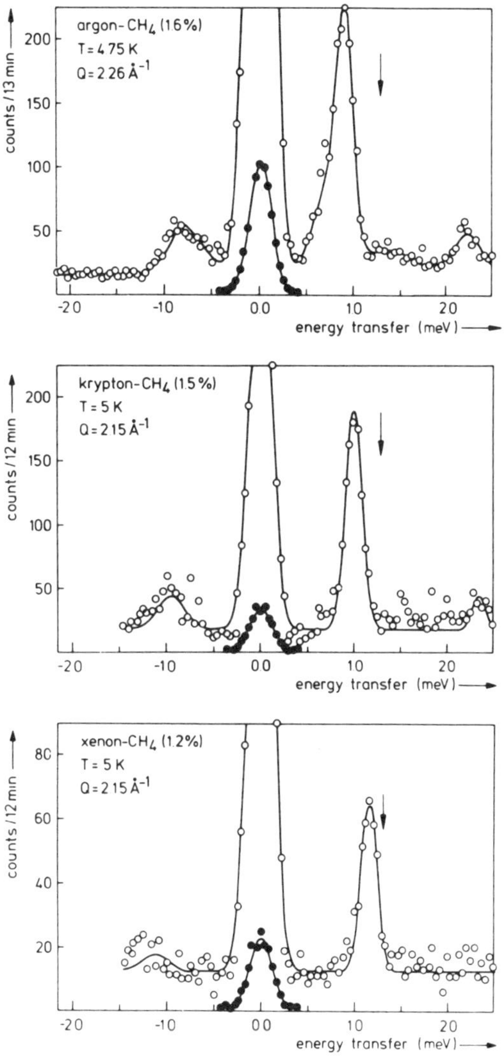

Fig.6.5. Inelastic neutron scattering and low-temperature rotational excitations of CH_4 impurities in solid rare-gas matrices /5.15/. The arrows mark the energy E = 2B, that is the J-0 to 1 transition for a free rotor. Solid lines refer to computer fits; the elastic intensity (full points) is scaled down by a factor ten

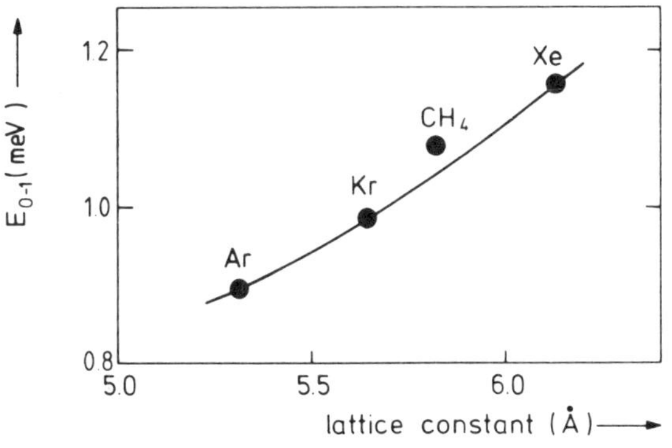

Fig.6.6. Energy of the perturbed J-0 to 1 transition of CH_4 plotted versus the lattice constant of the respective solid rare-gas matrix /5.15/

More elaborate attempts have been made /5.15/ to determine the potential parameters by calculating eigenvalues in a two-dimensional array spanned by the parameters V_4 and V_6 [$V_c(\omega^E)$ truncated at $\ell = 6$, free rotor functions with $J < 15$]. The set of parameters V_4 and V_6 best explaining all observations is selected. The results are not always unambiguous which means a unique set of parameters V_4 and V_6 cannot always be determined. This is quite unsatisfactory in view of the further aim of determining intermolecular interactions. Other problems may be related to the "tunneling between inequivalent sites", in which case orientation-dependent distortions of the lattice are suggested /6.11/. Therefore an analysis based on static potentials might be inappropriate for site symmetries which are not a subground of the molecular symmetry. Earlier optical measurements, which yielded less precise information, are analyzed in /5.29/.

The continuous transition from almost free quantum-mechanical rotation at low temperatures to classical rotational diffusion at high temperatures may best be observed with an argon matrix. Argon does not absorb neutrons and scatters only weak-

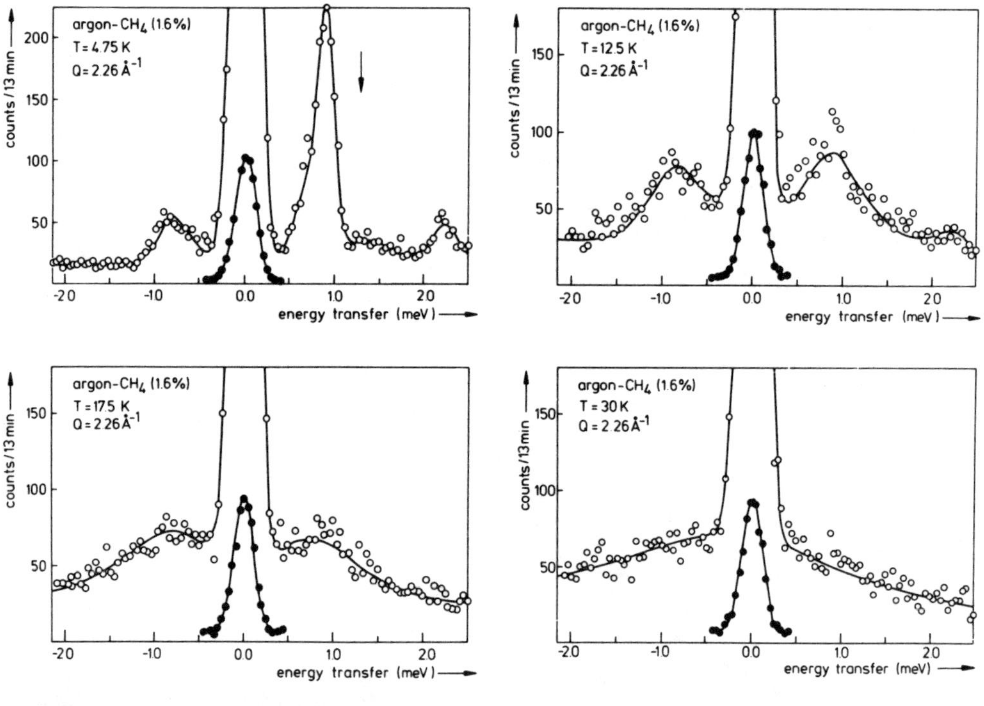

Fig.6.7. Inelastic neutron scattering spectra for 1.6% CH_4 in argon at different temperatures /5.15/. The data reflect the continuous transition from almost free rotation at low temperature to rotational diffusion above T ~ 30 K. Solid lines refer to computer fits; the elastic intensity is scaled down by a factor ten

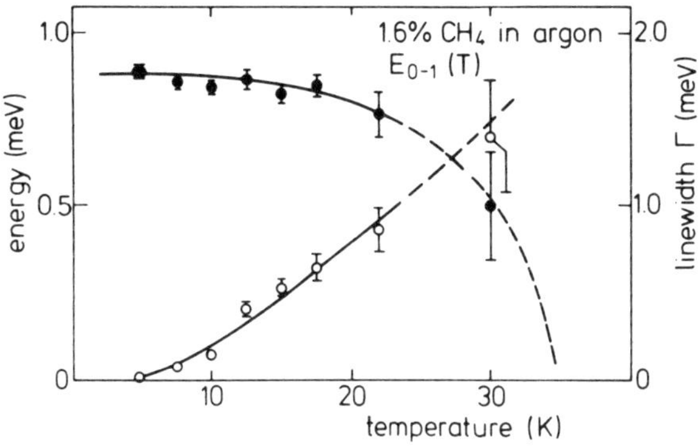

Fig.6.8. Temperature dependence of peak position and linewidth Γ (open points) of the perturbed J = 0 to 1 rotational transition of CH_4 molecules as substitutional impurities in an argon matrix. See also Fig.6.7 /5.15/

ly. Scans at different temperatures are shown in Fig.6.7. Rotational energies and linewidth for the 0-1 transition—as determined from least-squares fits to the spectra—are displayed in Fig.6.8. So far there is no quantitative explanation. Qualitatively the potential fluctuations V_{fl} increase with temperature when both rotational states and phonons are increasingly populated. One may add that V_{fl} is stronger in CH_4 for temperatures T > 20.4 K, that is, in the disordered phase; in addition to the phonons, the EOO interaction contributes to V_{fl} in phase I of bulk methane (CH_4I).

6.1.4 γ-Picoline

So far, there is just one known example of a CH_3 group rotating almost freely in a molecular crystal: γ-picoline (4-methyl-pyridine) /6.12,13/. The molecule consists of a pyridine ring and a methyl group opposite to the nitrogen atom (N⬡CH_3). Unfortunately the crystal structure is unknown. The neutron inelastic scattering experiments /6.13/ have been performed with a resolution at the elastic position ΔE = 0.25 meV (FWHM). A measurement at T = 5 K (Q = 1.5 Å^{-1}) with a polycrystalline sample (as in most reported experiments) is shown in Fig.6.9. Peaks at 0.52 ± 0.01 meV are interpreted as slightly perturbed 0-1 transitions. Much weaker additional peaks at 1.41 ± 0.03 meV and 1.92 ± 0.03 meV correspond to 1-2 and 0-2 transitions, respectively. This compares to free rotor levels at 0.665 meV (J = 1) and 2.62 meV (J = 2), with $E_J^{free} = BJ^2$.

The observations can be understood with a potential $V(\phi) = \frac{1}{2}V_6\cos6\phi$ with a barrier height $V_6 \cong 15 \pm 2$ meV (~ 180 K) /5.9,6.13/. A weak admixture of a cos3ϕ term cannot be ruled out, but probably is absent due to symmetry. This would be the case if the threefold axis of the CH_3 groups coincided with a twofold axis of the crystal. The periodic potential clearly is due to intermolecular interactions in the crystal, as the intramolecular barrier to CH_3 rotation is only 0.6 meV.

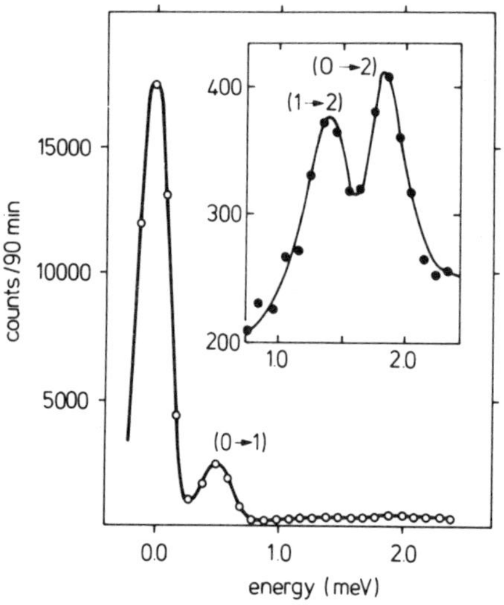

Fig.6.9. Neutron scattering spectrum of the rotational transitions of the methyl group in γ-picoline at T = 5 K /6.13/. The transitions are labelled by the corresponding free rotor transitions. In the insert the intensity scale is blown up in order to render the transitions to the level with J = 2 visible

6.2 Rotational Tunneling

One talks about rotational tunneling, if the rotational constant B is much smaller than the static potential ($B \ll V_{st}$) or, equivalently, if the scaled potential $V'_{st}(\omega) \ll 1$. In this case small ground-state splittings $\hbar\omega_0 \ll B$ result and the rotational wave functions more closely resemble oscillator functions than free rotator functions (Fig.5.2). For $V_{st} \gg B$ this may even be true in the presence of high symmetry and tunneling between nonequivalent sites.

Evidence of rotational tunneling in stoichiometric crystals was first deduced from low-temperature specific-heat anomalies (e.g., /6.14,15/) and residual entropies /5.25/. Usually only the high-temperature tail of a Schottky anomaly is observed. More recently NMR measurements also have provided access to rotational tunneling by 1) temperature-dependent measurements of the spin-lattice relaxation time T_1 /6.16/ (the method is very similar to an inelastic "fixed-window technique" /4.32/ and 2) level crossing techniques /6.17-20/. One modification of the latter method has been demonstrated with solid methane (CH_4); the levels of paramagnetic impurities (created by irradiation) have been tuned to resonance with the tunneling levels /6.17/. This measurement, which suffers somewhat from the perturbation caused by the impurities, has directly stimulated the first neutron measurement of rotational tunneling /5.5/.

In neutron measurements we are interested in systems with V_{st} only moderately larger than B. Otherwise the splittings of the librational ground-state are too small to be resolved (e.g., in NH_4I /6.20/). Measurements can successfully be performed for potentials $V'_{st} \lesssim 100$ B — that means $V \lesssim 750$ K for XH_3 and XH_4 groups (X = C,N). While the upper limit is given by the energy resolution of neutron scattering experiments (backscattering technique: $\Delta E \cong 0.3$ µeV) it is more or less a matter of taste where to place the limit between tunneling and almost free rotation. For orientational order (site symmetry ≦ molecular symmetry) $V'_{st} \gtrsim 30$ B is a possible choice. Splittings observed with neutrons range from rather large values $\hbar\omega_0 \cong B/10$ for CH_4 /5.5/ to $\hbar\omega_0 \cong B/500$ for DMA /4.32,6.21/ and $(NH_4)_2SnCl_6$ /4.27/. For splittings $\hbar\omega_0 < \Delta E$ the NMR techniques which recently have been developed are most appropriate and can complement neutron results. In the following, representative examples for the observation of rotational tunneling will be given for one- and three-dimensional rotors. Practically all measurements have been performed either at the reactor FRJ2 in Jülich and at the high-flux reactor of the ILL in Grenoble.

6.2.1 CH_3 Groups

a) Dimethylacetylene

Representative for the large class of systems with rotating CH_3 groups (which includes side groups in polymers) we shall first discuss dimethylacetylene (DMA). DMA (CH_3-C≡C-CH_3), methyl-substituted acetylene, is a relatively simple molecule. Two solid phases, both with a tetragonal cell, are known /6.22/. In the low-temperature phase the primitive cell contains two DMA molecules and the site symmetry ($C_1 \equiv 1$) is identical for all CH_3 groups. High-resolution measurements ($Q = 1.9$ Å^{-1}) at several temperatures are shown in Fig.6.10 /4.32/. They have been performed with polycrystalline samples.

At T = 4.5 K two inelastic peaks at ±1.7 µeV are found, located symmetrically around the elastic line, and represent transitions from A⇄E states (energy $\hbar\omega_0$). The tunnel splitting and an activation energy of 36 meV can be explained simultaneously with a potential $V(\phi) = \frac{1}{2}V_3\cos3\phi$ with $V_3 = 45 \pm 3$ meV. Higher order terms in the potential ($\frac{1}{2}V_6\cos6\phi$, etc.) seem to be negligible.

The above potential allows the prediction of an energy $E_1 = 15$ meV /6.17/ for the first librational state. Obviously, a measurement of E_1 would provide a test of the potential, predicted on the basis of the ground-state splitting.

Another interesting feature concerns the temperature dependence of the tunneling in DMA. It displays standard behavior in the sense that the lines shift towards smaller energies (above T ~ 20 K) and simultaneously broaden. A satisfactory fit /6.17/ of the temperature dependence of both the ground-state splitting $\hbar\omega_0$ and the linewidth Γ is provided by the stochastic averaging model of ALLEN /6.23/, which will be discussed in Sect.7.1.1. DMA is one of the few systems with one-dimensional

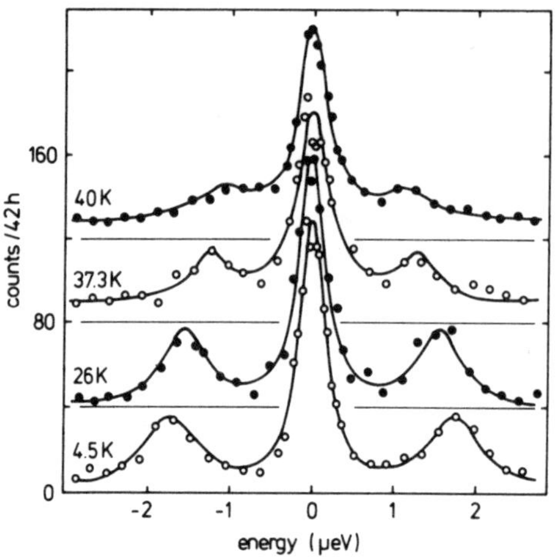

Fig.6.10. Inelastic neutron scattering spectrum from dimethylacetylene /4.32/ at several temperatures

rotations in which the structure is known and simple. Therefore DMA could well serve as a model system for further research.

b) MDBP

The most carefully studied system (but probably not the simplest one—its structure is not yet known) with CH_3 group tunneling is MDBP (4-methyl-2,6-ditertiarybutyl-phenol). A series of experiments employing NMR and inelastic neutron scattering has been performed. This has yielded detailed information both for samples with protonated molecules and for the partially deuterated compound MDBP (D_{21}) (deuteration except of CH_3 group). The observed tunnel splittings are rather large: 38 μeV for MDBP /6.24/ and 35 μeV for MDBP (D_{21}) /6.25,26/. Such relatively large splitting can still be determined by the backscattering technique. Different crystals have been used for the monochromator and the analyzer. On the other hand, the tunnel splitting of the first excited librational state (Fig.5.10) is sufficiently large to be observable with standard triple-axis methods. For MDBP (D_{21}) a splitting of 0.9 meV has been observed /6.27/, together with a mean librational energy $\overline{E}_1$ = 10.2 meV (Fig.6.11). In Sect.6.2.1c another successful measurement of an excited-state splitting will be reported. A number of facts support the assignment by CLOUGH et al. /6.27/: 1) the level scheme can consistently be explained with a potential $V(\phi) = -9.0\ \cos 3\phi + 1.8\ \cos 6\phi$ [meV], 2) the integrated intensities for the transitions to the two members of the doublet are the same, 3) partial deuteration has little effect on $\overline{E}_1$ and 4) the two peaks broaden with temperature in a way very similar to the broadening of the ground-state splitting. What is even more remarkable is the difference in width of the two peaks. While at low temperatures the width of the A state (that at higher energy) is resolution controlled, the E state

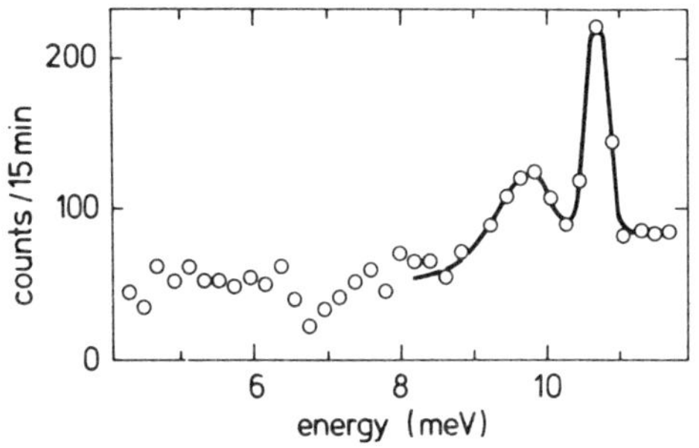

Fig.6.11. Inelastic neutron scattering spectrum of partially deuterated MDBP /6.27/. Two peaks at 9.75 meV and 10.65 meV are identified as first excited librational states with E and A symmetry, respectively. Apart from the excited-state tunnel splitting of 0.9 meV a broadening of the E level may be noted

is much broader and obviously has a shorter lifetime. This is explicable if only those potential fluctuations which modulate the amplitude and do not change the symmetry of the potential $V(\phi)$ are present. The ground-state wave functions as well as the perturbation $\mathscr{H}_{RL}$ (RL stands for rotation-lattice) then have A_1 symmetry. As the first excited librational state has A_2 symmetry, transition matrix elements vanish. In order to allow transitions, the phase of the potential needs to be modulated by lattice modes which gives rise to $\mathscr{H}_{RL} \sim \sin 3n\phi$. The unexpected broadening has important consequences concerning models for the temperature dependence (see Sect.7.1).

The temperature dependence of the tunnel splitting has also been measured for MDBP /6.25-28/. Two different activation energies $E_1^A = 5.6 \pm 0.8$ meV and $E_2^A = 11 \pm 1$ meV /6.28,29/ are required for a fit of the results. E_2^A seems to dominate above $T \gtrsim 15$ K; its value is very close to the energy of the librational levels. This is suggestive for the role of the librational states in the microscopic mechanism responsible for the broadening. A second process governed by E_1^A takes over at temperatures $T \lesssim 15$ K /6.28/. Here an excitation at 5.5 meV = E_1^A has been found in MDBP (D_{21}) and has been identified with an optical lattice mode. Its contribution in a second-order Raman process /6.28/ has been invoked. In addition to the broadening of the tunnel peaks, a quasielastic component has been found. Its width is always less than that of the tunnel peaks /6.25,26/. Both a direct inelastic measurement and the fixed-window technique have been applied to determine its temperature dependence.

On one hand, the observed features are of extreme importance in the explanation of the microscopic events responsible for the continuous transition from quantum-mechanical tunneling to classical reorientational motion. On the other hand, some "residual uncertainty" concerning the assignment of peaks in MDBP remains. Therefore experiments with other substances along similar lines as that with MDPB are highly desirable. It is clear that future efforts will aim in this direction. For further discussion of the temperature dependence of rotational tunneling based on MDBP results, see Sect.7.1.2.

c) Other Examples

There are many more examples with rotating CH_3 groups which all suffer from the fact that the crystal structures are not known or not sufficiently known. As one-dimensional rotation has the advantage of simplicity (e.g., rotational levels can be calculated exactly for certain potentials /5.8/), systematic crystallographic studies of these systems would be desirable. Below we shall give a short summary of several systems which may help to contribute to a better understanding of CH_3 rotation.

$Pb(CH_3)_4$ has already been mentioned in connection with the fixed-window method (Sect.4.5.1). Direct measurements at 3 K yield a tunnel splitting of 35 μeV /4.32/. Barely above 3 K the lines broaden significantly. Additional data could provide an interesting test of recent models pertaining to the temperature dependence of tunneling states (Sect.7.1).

A rather systematic effort has been made to investigate methyl-substituted pyridines (lutidines and picolines; see also Sect.6.1.4) /6.30-32/ as well as toluene /6.33/, both with NMR T_1 and neutron measurements. Rotational potentials have been determined [neglecting contributions V_{3n} with $n \geq 3$ and the phases ϕ_{3n} in $V(\phi) = \frac{1}{2} \sum V_{3n} \cos(3n\phi + \phi_{3n})$] from neutron measurements of the tunnel splitting (A-E transition) and of the librational excitations. The comparison with results based on NMR data for the same compound is not always satisfactory. The temperature dependence of the relaxation time T_1 is analyzed in terms of a classical model for the high-temperature relaxation and a semiquantum mechanical model /6.34/ for the low-temperature relaxation. The method seems to yield fairly reliable potentials if the measurements are extended far into the regimes of low-temperature and high-temperature relaxation /6.31/. As this is not always possible, rotational potentials based on NMR T_1 data alone appear to be rather uncertain.

2,6-Dimethylpyridine has a large tunnel splitting with $\hbar\omega_0$ = 190 μeV and therefore represents another successful measurement of a splitting of the first librational state. Two peaks at E_{1E} = 3.6 meV and E_{1A} = 4.3 meV have been observed /6.31/.

Another very interesting system is provided by $SnF_2(CH_3)_2$ /6.35,36/. Only the structure of the (body-centred) tetragonal high-temperature phase is known. In this phase the CH_3 groups are arranged in planes and their rotation axes form a square lattice; the square lattice probably becomes (weakly) distorted below the phase transition at T_0 = 70 K. From other experimental evidence, tunneling states were predicted at about 15 μeV /6.35/ and actually were found at 13.7 μeV /6.36/. $SnF_2(CH_3)_2$ appears to be an excellent candidate for further measurements, in particular of $\underline{Q}$-dependent intensities (with single crystals) aimed to learn about rotational wave functions.

6.2.2 NH_3 (in Hexamine Nickel Halides)

Usually tunneling spectroscopy by neutron scattering is complemented by NMR measurements (or vice versa). For systems $Ni(NH_3)_6X_2$ with X = I, Br or Cl this is different. Evidence for low-energy rotational phenomena was first obtained in specific-heat measurements, down to temperatures of about 0.1 K, by van KEMPEN et al. /6.14/. In all three compounds these authors observed Schottky anomalies which they attributed to a two-level system with a small energy separation $\Delta(X)$. Values $\Delta(I)$ = 62 μeV, $\Delta(Br)$ = 10.5 μeV and $\Delta(Cl)$ = 3-4 μeV have been extracted from the data. Because the splittings were small, only high-temperature tails of the anomalies $C \cong (1/4)k(\Delta/T)^2$ could be observed for the bromide and the chloride. Therefore the values $\Delta(Br)$ and particularly $\Delta(Cl)$ appeared to be rather uncertain. A quantitative explanation in terms of a tunnel splitting $\hbar\omega_0$ of the NH_3 group has been given by BATES and STEVENS /6.37/ on the basis of an electrostatic model. Qualitatively, the rotational potential decreases when the lattice parameter and the intermolecular distances increase with the ionic radius of the respective halide ion (from Cl^- to I^-). A similar trend may be noted for the phase-transition temperatures $T_0(I)$ = 20 K, $T_0(Br)$ = 45 K and $T_0(Cl)$ = 76 K.

The structure of the high-temperature phase has already been discussed in Sect. 4.3.1 in connection with the one-dimensional continuous rotational diffusion in $Ni(NH_3)_6I_2$. It is shown in Fig.4.1. The alignment of the dipole moment of the NH_3 molecules in the crystal is responsible for uniaxial rotations around the threefold axis of the molecule. All three salts display orientational order-disorder transitions to an ordered low-temperature phase (which is not the same for the three salts) in which rotational tunneling is observed. The details of the low-temperature structure have not yet been completely established for any of the three compounds /2.25,6.37/. Probably there are two inequivalent sites with NH_3 molecules in the unit cell. This introduces some ambiguity into the interpretation of low-temperature measurements.

Obviously there is an interest in testing the specific-heat results with direct neutron measurements. For $Ni(NH_3)_6I_2$ an energy resolution of only intermediate quality is required and therefore a three-axis spectrometer with a resolution $\Delta E \cong 50$ μeV has been used. A tunnel splitting $\hbar\omega_0(I) = 63 \pm 2$ μeV has been found, in excellent agreement with the corresponding specific-heat value. In addition a rather broad peak at 9.0 meV has been observed and assigned to a transition to the first excited librational peak /2.25/. Maybe the observed "broadening" of this peak is due to an unresolved splitting of the energy E_1 into E_{1A} and E_{1E} (Fig.5.10); this question deserves further experimental efforts.

For the remaining two compounds the measurements have been performed with a backscattering spectrometer and a resolution of 0.36 μeV. A ground-state splitting $\hbar\omega_0(Br) = 8.0 \pm 0.3$ μeV is found /6.38/—still in reasonable agreement with the specific-heat value of 10.5 μeV. The tunnel splitting of the NH_3 groups in $Ni(NH_3)_6Cl_2$ cannot be resolved. Thus it is much smaller than the corresponding value $\Delta(Cl)$, obtained from specific-heat measurements. A broadening of the elastic line is observed, however, and it may serve for an estimate of a splitting $\hbar\omega_0(Cl) \cong 0.1$ μeV. For a discussion of the rotational potential for the three salts, one is referred to /6.38/.

6.2.3 Methane

a) CH_4 II

As has already been mentioned, the first neutron measurement of rotational tunneling has been performed in phase II of CH_4 /5.5/. The structure of this phase is shown in Fig.6.1. In Sect.6.1.2 we have discussed neutron spectra displaying the excitations of the disordered molecules (2 molecules out of 8 in the unit cell). These molecules only experience the crystal field $V_c(\tau)$. We now turn to the ordered molecules (6 molecules out of 8) where the potential consists of two contributions: 1) the crystal field with octahedral symmetry and 2) a molecular field $V_M(\tau)$ with tetrahedral symmetry due to the octopole-octopole interaction. Measurements performed with a three-axis spectrometer and an energy resolution $\Delta E \cong 40$ μeV are shown in Fig.6.12 /5.5/. Two inelastic peaks are observed, both on the energy-gain and the energy-loss side of the spectrum. The observed splittings $\hbar\omega_{01} = 143 \pm 3$ μeV and $\hbar\omega_{02} = 73 \pm 3$ μeV correspond to A-E and T-E transitions, respectively—with $\hbar\omega_{01} = E_T - E_A$ and $\hbar\omega_{02} = E_E - E_T$. The absence of an A-E transition was originally explained by nuclear spin conservation /5.5/. Such a transition requires a change of the total nuclear spin of a methane molecule from I = 2 (A state) to I = 0 (E state), as can be seen in Fig.5.11. Changes $\Delta I = \pm 2$ cannot be caused by a scattered neutron

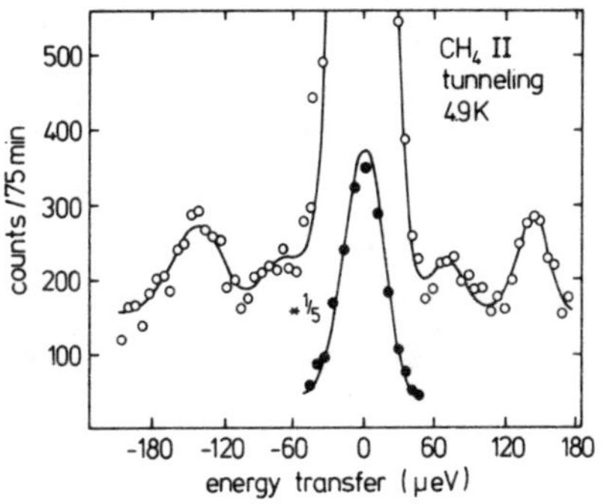

Fig.6.12. Tunneling lines in CH_4 II /5.5/. The solid lines represent a fit to the data; the intensity of the elastic line is scaled down by a factor 5

and therefore the transition is forbidden. The same argument holds for NH_4^+ ions. A more general argument for this selection rule has been based on the symmetry of the neutron scattering operator (Sect.5.3).

A ratio $\omega_{01}/\omega_{02} \cong 2$ of the observed energies is found, which indicates that the 180° overlap is indeed very small compared to the 120° overlap (5.3) (Table 5.4). This is also the reason why no T state splitting is observed, in spite of the site symmetry $\bar{4}2m$. In principle such a splitting originates from the inequivalence of the 180° matrix elements H_x and H_z. It has been estimated to be of the order of 0.1 μeV /5.9/. Detailed analyses of the level scheme as a function of the rotational potential may be found in /5.9,31/. HÜLLER and KROLL /5.9/, for example, studied the influence of the crystal field by taking a potential $V(\tau) = V_M(\tau) + V_c(\tau)$ with $V_M(\tau) = A_3H_{11}^{(3)}(\tau)$ and $V_c(\tau) = A_4\tilde{H}_{11}^{(4)}(\tau)$. The tilde denotes that $V(\tau)$ is expressed in a rotated frame in which the equilibrium orientation of the CH_4 molecule is a standard orientation (Fig.5.6). In this frame, $V_M(\tau)$ has the simple form given above, while $V_c(\tau)$ cannot be expressed in terms of just one cubic rotator function $H_{11}^{(\ell)}(\tau)$ of order ℓ.

The intensities of the observed peaks have recently been compared to theoretical results, based on the calculation of transition matrix elements /5.35/. Also, a strong shift of the tunneling states has been observed, when changing to deuterated methane (CD_4) /5.39/. This isotope effect will be discussed in Sect.7.3.

Press and Kollmar also investigated the temperature dependence of the rotational tunneling in CH_4. Results, obtained with a slightly relaxed energy resolution are summarized in Fig.6.13. The width displays the usual increase with temperature. The tunneling frequencies, however, show an unusual behavior — they increase with temperature. This is due to the orientational order-disorder transition at $T_0 = 20.4$ K. The orientational order parameter and hence also the molecular field (magnitude A_3) decreases on approaching the phase transition. The orientational order parameter is directly related to the intensity of superlattice peaks which have been observed as a function of temperature in CD_4 /2.24,6.39/. If one forgets the fluctuating part V_{fl} of the potential for the moment, the decrease of the potential leads to an increased overlap of the wave functions and, therefore, to a larger tunnel splitting. A calculation yields qualitative agreement with the experiment /2.22/. The inclusion of the fluctuating part of the potential V_{fl} would lead to a smaller increase of the splitting. Above T_0 quasielastic scattering caused by (almost) continuous rotational diffusion of all molecules in CH_4 I is observed /5.4,5/.

Another interesting phenomenon concerns the tunnel splitting at low temperatures. An increase of E_{AT} and E_{TE} with decreasing temperature has been predicted /2.22/. This is due to the fact that the magnitude of the octopole-octopole interaction de-

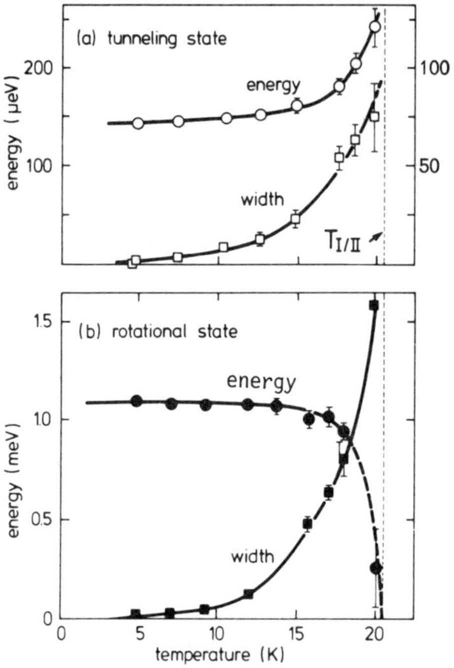

Fig.6.13a,b. Temperature dependence of the energy and the linewidth of (a) the A-T tunneling transition and (b) the 0-1 rotational transition in CH_4 II /5.5/. The right scale in (a) refers to the linewidth

pends on the spin species. At low temperatures A states become increasingly populated. Their rotational wave function is symmetric and therefore is positive everywhere. The T and E state wave functions possess zeros. In order to compensate this effect the maxima of the T and E state wave functions are more strongly peaked. The angle dependence of the A state is smoother, it has a "more spherical" density distribution and hence a smaller effective octopole moment than the other spin species. Recently a neutron scattering experiment was performed in the temperature range $1.4\ K \leq T \leq 5\ K$ /6.40/ in order to test the theoretical predictions. An energy resolution of about 10 μeV was employed. While the absolute values for the ground-state splittings are not well reproduced by the theory /2.22/ (this was already clear from the earlier measurements /5.5/), the temperature dependence predicted by the theory fits rather well.

b) CH_4 Adsorbed on Grafoil

Recently the tunneling states of CH_4 on graphite surfaces have been observed /6.41, 42/. Methane molecules adsorbed on grafoil form a triangular two-dimensional lattice. For CH_4 coverages of 0.9 monolayers or less, a commensurate $\sqrt{3}*\sqrt{3}$ structure is observed. $\sqrt{3}*\sqrt{3}$ means that the lattice constant of the two-dimensional methane lattice is $\sqrt{3}$ times the lattice constant of graphite. The distance of the molecular c.o.m.

from the surface is 3.3 Å. For higher coverages a change to a more compressed phase, which is not in registry with the substrate, takes place /6.41/.

The Oxford group /6.42/ observes five lines at 17 μeV ($T_{1,2} \rightarrow T_3$), 39 μeV ($T_3 \rightarrow E$), 56 μeV ($T_{1,2} \rightarrow E$), 94 μeV ($A \rightarrow T_{1,2}$) and 112 μeV ($A \rightarrow T_3$) (Fig.6.14). The respective assignment is given in brackets. Spectra have been recorded with two different resolutions: 1) time-of-flight spectrometer, ΔE = 20 μeV and 2) backscattering, ΔE = 1 μeV. The energies listed above are accurate to within 2 μeV. The results have been complemented by measurements of excited librational states.

The magnitude of the observed tunnel splittings is very similar to that found in bulk methane /5.5/. Taking the average T state energy $\bar{E}_T = (1/3)(2\,E_{T_{1,2}} + E_{T_3})$ as a reference energy, the values E_{AE} and E_{ET} are about 25% smaller than in CH_4 II. The observed 5 lines can consistently be explained by the presence of a potential with trigonal symmetry (Table 5.4). There is a site with trigonal symmetry in the basal plane of graphite, namely, the position of the carbon atoms. It is possible that the CH_4 molecules sit above such sites in the registered phase and thus retain a threefold axis.

Two kinds of interacting contribute to the rotational potential: 1) an interaction CH_4 substrate and 2) a direct interaction between CH_4 molecules. The authors conclude that the second contribution is more important. Calculations of the ground-state splittings have been based on phenomenological pair potentials /6.42/.

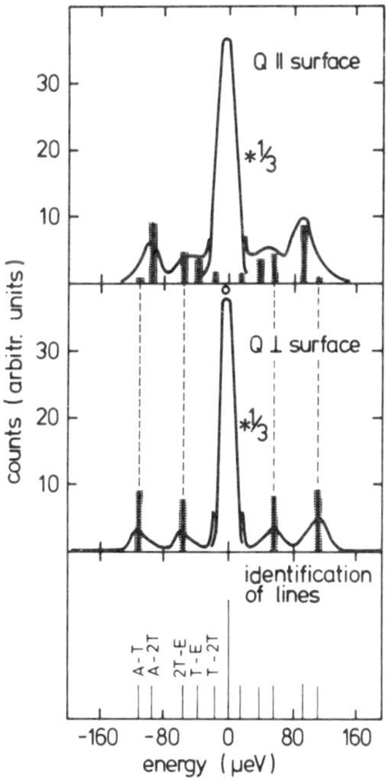

Fig.6.14. Tunneling spectrum of CH_4 adsorbed on Vulcan II /6.42/ for a momentum transfer $Q_\parallel$ and $Q_\perp$ (with respect to the hexagonal a-b plane). The difference of the intensities can be explained by a calculation of neutron scattering transition matrix elements /3.4/ with h_1 = -14 and h_4 = -9.1. Smoothened experimental data are represented by a line

A very interesting feature concerns the intensities observed with a substrate of exfoliated graphite. In such a sample the hexagonal c axis is largely ordered, whereas there is no order referring to the a and b axis. The neutron spectra differ significantly for $\underline{Q}$ parallel ($Q_{\parallel}$) and perpendicular ($Q_{\perp}$) to the basal plane. Inspection of transition matrix elements for neutron scattering /3.3,4/ shows that a significant contribution to the intensities $I(Q_{\perp})$ originates only from the A-T_3 and $T_{1,2}$-E transitions, when assuming trigonal site symmetry for the methane molecules. This in accord with experimental observations and provides an example for intensity information supporting the assignment of rotational levels (Fig.6.14).

6.2.4 Ammonium Salts

a) $(NH_4)_2SnCl_6$

The face-centred cubic structure of $(NH_4)_2SnCl_2$ is shown in Fig.4.7. Rotational excitations in $(NH_4)_2SnCl_6$ already have been discussed in Sect.4.5.1 in connection with the classical high-temperature reorientational motion of the NH_4^+ groups. We now turn to the tunneling states of this compound as observed with a backscattering spectrometer and a resolution of 0.38 μeV. First the results at low temperatures (T = 6 K; Fig.6.15) shall be discussed. As for the ordered molecules in CH_4 II, two pairs of inelastic peaks are observed: $\hbar\omega_{AT}$ = 2.96 ± 0.04 μeV and $\hbar\omega_{TE}$ = 1.51 ± 0.03 μeV /4.27/. The selection rule responsible for the absence of an A-E transition has been discussed before (Sects.5.3, 6.2.3). The ratio ω_{AT}/ω_{TE} again is 2 within error bars, thus indicating the dominance of the 120° overlap. Excitations at 13.4 meV and 30 meV help to complete the picture /4.27/. They are inter-

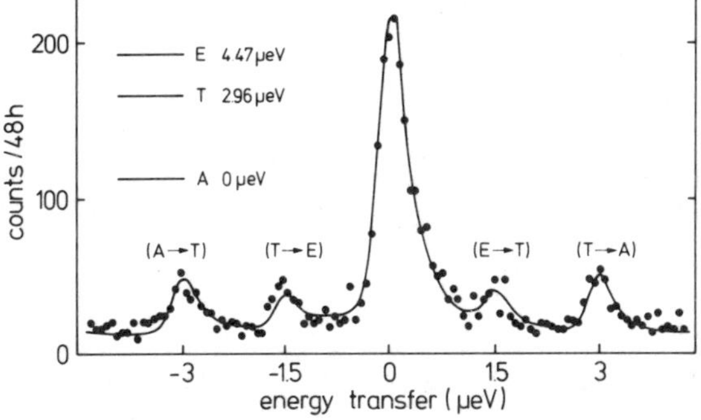

Fig.6.15. Inelastic neutron scattering spectrum of $(NH_4)_2SnCl_6$ measured with a backscattering spectrometer /4.27/. The insert shows the level scheme of the librational ground-state of the NH_4^+ group

preted in terms of transitions to the first and second excited librational states and signal the presence of a boxlike potential. Attempts have been made to explain all observed features with a rotational potential of tetrahedral symmetry (2.21) /4.27,5.28/, so far with limited success.

Additional tunnel splittings have recently been observed by PUNKKINEN et al. /6.43/ with NMR methods at 0.05_4, 0.10 and 0.15_3 µeV. These findings are interpreted in terms of transitions within the T state multiplet. According to Punkkinen et al., the T state degeneracy is completely removed because the site symmetry is lower than tetrahedral (from Table 5.4: site symmetry m or 1, if all NH_4^+ sites are equivalent). This indicates a weak structural distortion (phase transition?) which so far has escaped crystallographic studies. Additional experiments are required to clarify the situation.

HÜLLER has calculated the intensities of the observed inelastic lines by performing powder averages (3.19) of the double differential neutron scattering cross sections /3.3/. The elastic lines have not been included in the comparison because of additional uncertainties concerning scattering from the sample container, incoherent scattering from other atoms in the sample, etc. A comparison between experimentally observed intensities for both $(NH_4)_2SnCl_6$ /4.27/ and NH_4ClO_4 /6.44,45/ and theoretical results is shown in Fig.6.16 and Table 6.1. For $(NH_4)_2SnCl_6$ the calculations are based on tetrahedral site symmetry. Excellent agreement is found.

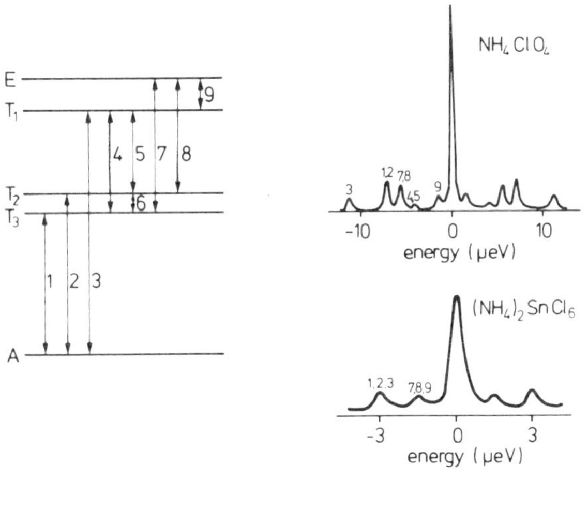

Fig.6.16. Level scheme of a regular tetrahedron in a potential which has no symmetry. This is compared to schematically drawn neutron scattering spectra for NH_4ClO_4 /6.44,45/ and $(NH_4)_2SnCl_6$ /4.27/, which represent examples with low and high site symmetry /3.3/

Table 6.1. Energies and intensities (arbitrary units) of the tunneling transitions in NH_4ClO_4 /6.45/ and $(NH_4)_2SnCl_6$ /4.27/. The experimentally determined intensities are compared to theoretical values /3.3/ which are given in units of $(1/24)Na^2_{inc} \cdot j_0(Q_0\rho)$

	NH_4ClO_4					$(NH_4)_2SnCl_6$	
Line	1,2	3	4,5	7,8	9	1,2,3	7,8,9
$\hbar\omega$ (μeV)	7.17	11.28	4.11	5.65	1.51	2.96	1.51
intensity (exp.)	104	49	22	84	50	35	26
intensity (calc.)	10	5	2	8	4	15	12

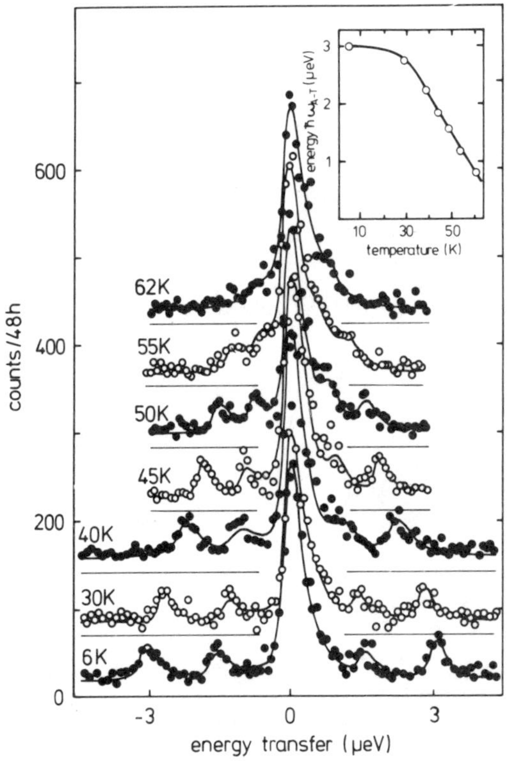

Fig.6.17. Inelastic neutron scattering spectra of $(NH_4)_2SnCl_6$ at various temperatures. The insert shows the temperature dependence of the outer line ($\hbar\omega_{AT}$) as obtained from fits to the data /4.27/

The temperature dependence of the tunneling lines in $(NH_4)_2SnCl_6$ is shown in Fig.6.17. As usual the lines shift towards lower energies. They do not reveal significant broadening, however. Therefore tunnel splittings remain visible up to a temperature T ≅ 60 K, and it appears that the classical high temperature regime is only reached above temperatures T ≅ 70 K. As the ratio $\omega_{AT}/\omega_{TE} \cong 2$ seems to persist as well, the data have been fitted with the constraint $\omega_{AT}(T) = 2\,\omega_{TE}(T)$ (Fig.6.17).

It is possible that the absence of broadening is related to the short lifetime of the first excited librational state (random averaging model, see Sect.7.1.1), which even decreases with temperature [Γ(T = 6 K) ≅ 1 meV].

b) NH_4ClO_4

Ammonium perchlorate (NH_4ClO_4) has an orthorhombic structure with four molecules per unit cell /6.46/. All molecules are structurally equivalent, that is, they experience the same static rotational potential. The only symmetry element at a NH_4^+ site in NH_4ClO_4 is a mirror plane /6.46/. No phase transition has been reported for T < 300 K. From earlier NMR T_1 measurements a tunnel splitting of about 2 µeV has been concluded /6.47/ and a constraint to uniaxial rotation has been discussed as a possible consequence of hydrogen bonding /6.43/.

Neutron spectra recorded at low temperatures /6.44,45/ rule out the latter possibility and show that conventional T_1 measurements can provide a result in the correct frequency range, but cannot reveal the detailed level scheme. The tunneling spectrum (Fig.6.18) looks rather complex at first glance. Five transitions are observed for energy transfers $|E| \lesssim 12$ µeV; the energies are listed in Table 6.1. An extension of the measurements beyond 12 µeV did not yield additional lines. There were some initial difficulties in constructing a level scheme from the observed transitions. Finally an explanation was based on the known site symmetry, but needed the inclusion of an accidental degeneracy of two of the three T levels. As may be seen from Table 5.4 or Fig.5.7, this is equivalent to assuming an effective site symmetry, 3 or 3m. These symmetries give rise to relations between the overlap matrix elements which necessitate the observed degeneracy.

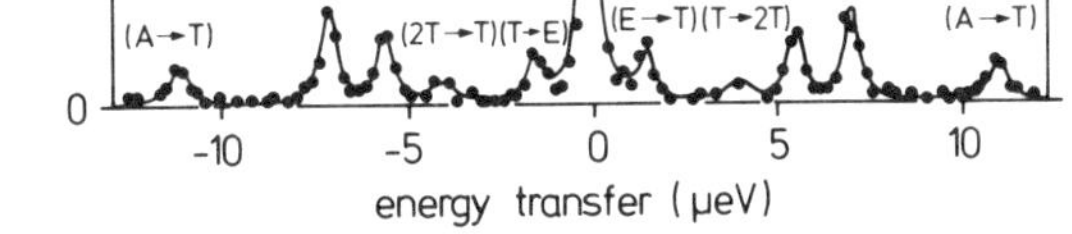

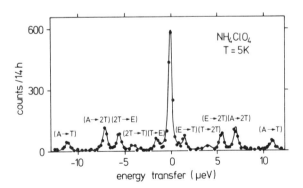

Fig.6.18. Inelastic neutron scattering spectrum of NH_4ClO_4 (T = 5 K) measured with a backscattering spectrometer /6.45/. Note that the time indicated refers to the whole spectrum

HÜLLER /3.3/ has fitted the observed tunneling energies (neglecting the 180° overlap) and finds h_1 = -0.038 μeV and $h_2 = h_3 = h_4$ = -1.410 μeV. This corresponds to three easy axes and one hard axis of rotation. Thus the rotational potential is clearly three-dimensional. Unfortunately, attempts to calculate V(τ) on the basis of electrostatic interactions remained unsuccessful. The fact that the threefold rotation axis connected with the matrix element h_1 is a "hard axis" (therefore little overlap) also is reflected in the librational amplitudes which have been found in the structure analysis /3.3,6.45/. In a NMR level crossing experiment /6.48/ a splitting of ≅ 0.18 μeV has been observed. This probably represents the T state splitting which originally could not be resolved by neutron scattering /6.44,45/ and lead to the assumption of an accidental degeneracy /3.3/. A comparison between measured and calculated neutron intensities is shown in Table 6.1. Both, calculation /3.3/ and experiment /6.44,45/ refer to powder samples and, as in the case of $(NH_4)_2SnCl_6$, very good agreement is found.

Recently, the neutron spectra have been reanalyzed /6.49/ using the 120° overlap matrix elements h_i directly as fit parameters. This has become possible since energies and intensities of the transitions can be calculated directly (/3.4/ and appendix) starting from a general set of h_i. As a result matrix elements h_1 = -0.035 ± 0.005 μeV, $h_2 = h_3$ = -1.317 ± 0.004 μeV and h_4 = -1.410 ± 0.006 μeV are found, which yields a splitting $E_{T_1} - E_{T_2}$ = 0.24 ± 0.02 μeV. The agreement with the NMR result is surprisingly good. We may conclude that the inclusion of intensities into the fit is superior to the conventional analysis and yields more detailed information, as long as the width of the wave functions can be neglected.

The temperature dependence of the transitions within the ground-state multiplet of NH_4ClO_4 has been measured and analyzed as well /6.45/. However, due to the complexity of the spectra, NH_4ClO_4 does not provide a model example suitable for comparison with current theories. It may be noted that the observed broadening of the lines seems to show an activation behavior with an activation energy E^A = 22 meV. The meaning of this value is not quite clear; the first librational state is found at a much lower energy /6.45/ and does not seem to be related to E^A.

7. Rotational Excitations at Low Temperatures III. Special Features

Several particularly interesting subjects have already been briefly mentioned either in the general discussion of low-temperature rotations in Chap.5 or in context of the example presented in Chap.6. They include the continuous transition from low-temperature to high-temperature rotation, the isotope effect and the pressure dependence of the rotational states. It appears useful to discuss these and related aspects in more detail in a final chapter.

7.1 Temperature Dependence

One of the most fascinating aspects of single particle rotation is the continuous transiton from quantum-mechanical rotation at low temperatures to classical diffusive rotational motion at high temperatures. In neutron scattering this means a transition from line spectra at low temperatures to quasielastic scattering at high temperatures. In Chap.6 we have given a number of examples for this temperature dependence, both close to the limit of free rotation and in the limit of rotational tunneling. The common feature of all examples (CH_4 II is an exception, because of its order-disorder phase transition) is a shift of the peak position(s) of the tunneling line(s) towards lower energies, before they merge into a quasielastic line at high temperatures. In most cases this shift is accompanied by a broadening of the lines.

It appears that access to the mechanism behind this temperature dependence is of extreme importance for a final understanding of single particle rotations. Because of the importance of the topic, numerous attempts have been made to solve this problem. We may note, however, that in spite of considerable progress and some success of current theories, the final aim has not yet been reached. This is partly due to the fact that rather often standard information about the investigated substances has not been available concerning 1) static aspects: the structural parameters and 2) dynamical aspects: the conventional lattice dynamics. Furthermore mostly powder samples have been used in the experiments. Apparently the lattice dynamical excitations, which become increasingly populated with temperature, give rise to a fluctuating potential and are responsible for the observed phenomenon. The models put forward either need not specify the nature of these excitations at all /6.23,25/ or base the explanation on either acoustic phonons /6.11/, translational optic modes /6.26,27/, librational modes of the crystal /7.1/ or internal modes of the molecules /7.2/. Below, a brief review of these models will be given. All calculations deal with the simpler case of uniaxial rotation of CH_3 groups. An extension to three-dimensional rotational motion so far has not been attempted.

7.1.1 Random Averaging Model

The most frequently applied—but also least specific—model has been coined by ALLEN /6.23/. In this model the temperature dependence is explained in terms of a dynamical averaging between two frequencies—the tunnel splitting of the ground state ω_0 and of the first excited state ω_1. The averaging is due to thermally activated transitions between these states, the tunnel splittings of which have opposite signs (Fig.5.10). The magnitude of the splittings can be calculated exactly by solving Mathieu's equation. Allen has adapted a semiclassical model of ANDERSON /7.3/ (later reviewed in /7.4/) which describes the frequency spectrum of an oscillator which switches randomly between two frequencies. This switching is treated as a stationary Markov process /7.8/. While Anderson gave an example with two frequencies $\pm\Delta$ and equal populations p_i and lifetimes w_i^{-1} of the states, Allen took the values pertaining to a rigid triangular group rotating in a potential $\frac{1}{2}V_3\cos 3\phi$. The general expression for the spectrum function $I(\omega)$ /7.3/ reads

$$I(\omega) = \mathrm{Re}\,\{\underline{p}\,A\,\underline{1}\} \tag{7.1}$$

with $\underline{1} = (1,1)$. In case of a tunneling CH_3 group /6.23/ $\underline{p} = (p_0,p_1)$ denotes the population of the librational states (Boltzmann statistics) and $A = A_1+A_2$ is a matrix composed of a diagonal matrix containing the discrete frequencies of the rotor

$$A_1 = i\begin{pmatrix} (\omega_0-\omega) & 0 \\ 0 & (\omega_1-\omega) \end{pmatrix} \tag{7.2}$$

and the matrix of relaxation rates

$$A_2 = \begin{pmatrix} -w_0 & w_0 \\ w_1 & w_1 \end{pmatrix} \tag{7.3}$$

w_0 represents the rate of transitions from the ground state and, in an analogous fashion, w_1 the rate of transitions from the excited state. It is assumed that the transition rates are related by $w_0 = w_1 \exp(-E_1/kT)$. E_1 is introduced as a mean librational frequency (geometrical average) which needs to be defined for larger tunnel splittings.

The evaluation of (7.1) yields a spectral function $I(\omega)$

$$I(\omega) = \frac{w_1(\omega_0-\omega_1)^2 \exp(-E_1/kT)\,[1+\exp(-E_1/kT)]^{-1}}{[(\omega-\omega_0)(\omega-\omega_1)]^2 + w_1^2[(\omega-\omega_0)+(\omega-\omega_1)\exp(-E_1/kT)]^2} \quad . \tag{7.4}$$

In discussing the above expression it is important to note that the tunnel splitting is much larger in the excited state and has the opposite sign (Fig.5.10). The magnitude of the relaxation w_1 (excited state) determines the relative importance of the two terms in the denominator. For small w_1 and low temperatures the temperature-independent first term dominates and gives rise to two discrete tunneling frequencies close to ω_0 and ω_1.

The second term in the denominator increases with temperature and tends to establish an average frequency close to $\omega_p = \omega_0+\omega_1 \exp(-E_1/kT)$ which becomes smaller at higher temperatures. In order to maintain a non-negative value for the peak position, librational states with $n > 1$ need to be included. For large relaxation rates w_1, a rather loosely defined generalization for the peak positions $\omega_p(T)$ reads /7.5/

$$\omega_p(T) = \sum_{n=0}^{N} M_n\omega_n(T=0)\exp(-E_n/kT) \Big/ \sum_{n=0}^{N} M_n \exp(-E_n/kT) \quad . \tag{7.5}$$

Multiplicities M_n have been included in order to extend the applicability of the above expression to three-dimensional rotors. The equivalent expression for the energy width is given in /7.5/.

For low temperatures, one has $\exp(-E_1/kT) \ll 1$ and for frequencies $\omega \leqq \omega_0$ the denominator of (7.4) can be approximated /6.25/ by $(\omega_0-\omega_1)^2 + w_1^2[(\omega-\omega_0-\delta_p)^2 + a^2]$. The following definitions have been introduced:

$$\delta_p = -(\omega_0-\omega_1)\exp(-E_1/kT)/(1+x^2) \tag{7.6a}$$

$$a = w_1 x^2(1+x^2)^{-1}\exp(-E_1/kT) = \delta\cdot x \tag{7.6b}$$

$$x = -(\omega_0-\omega_1)/w_1 \quad . \tag{7.6c}$$

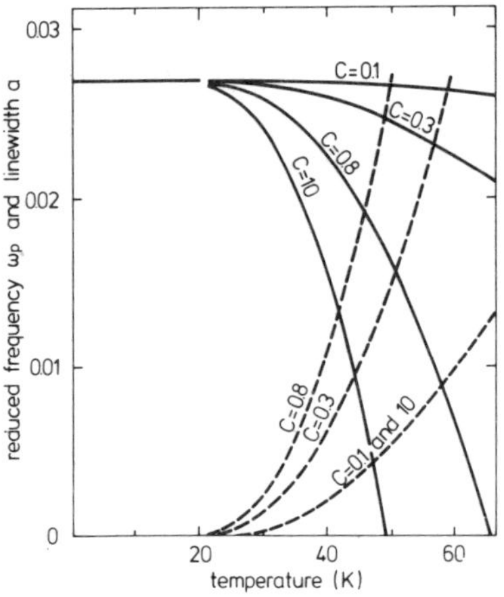

Fig.7.1. Temperature-dependent position and width of the inelastic lines of a tunneling CH_3 group /6.23/. For the reduced jump rate w_1/ω_1 an exponential form $w_1/\omega_1 = c\exp(-E_1/kT)$ is assumed. The solutions of Mathieu's equation for a CH_3 group in a potential of magnitude $V_3 \cong 55$ meV are used, and results are given for several values of the parameter c. Note the little broadening in case of fast relaxation (c = 10)

The spectral function $I(\omega)$ then has a Lorentzian shape. It is centered at $\omega_p = \omega_0+\delta_p$ (δ_p is negative!) and has a width (HWHM) α.

Examples both for the peak position $\omega_p(T)$ and the width α are given in Fig.7.1 /6.23/ on the basis of (7.4). The solutions of the Mathieu equation for a potential V_3 = 55 meV are used, and the relaxation rate w_1 serves as a parameter. Usually a strong shift δ_p is accompanied by a broadening of the same magnitude. If, however, the temperature-dependent term in the denominator of (7.4) is strongly weighted due to a large w_1, the change of the width α [see definition in (7.6b)] remains relatively small. An example where this phenomenon as well as a short lifetime w_1^{-1} of the excited state has been observed is $(NH_4)_2SnCl_6$ /4.27/. The relaxation rate w_1 enters Allen's model of the temperature dependence in addition to the parameters which describe the tunneling at low temperatures. w_1 can be determined by fits to the observed temperature dependence and then can be compared to direct measurements of w_1.

Generalization of Allen's model to the tunneling of tetrahedral molecules /4.27, 7.5/ so far are somewhat questionable as they do not account for the more complicated level scheme. One aspect of the random averaging model is that only differences between the energies of the A and E states enter (Fig.5.10). This can be made plausible for strong potentials with very small ground-state splitting. Then a pocket state can be chosen as an eigenstate which can be written as a linear combination of symmetrized states. During excursions to the excited state, the A and E states dephase with the difference of their librational state energies. This difference is the tunnel splitting ω_1 of the first excited state.

7.1.2 Refined Random Averaging Model

A combined effort of both testing Allen's model with NMR *and* neutron measurements (particularly with MDBP, see Sect.6.2.1b) and incorporating additional aspects into the model published by ALLEN /6.23/ has been made by CLOUGH and collaborators /6.27, 28/. In interpreting Allen's model they distinguished between two different "relaxation mechanisms". Their meanings are discussed in connection with the evolution of a pocket state, say, Φ_1 (with proton 1 in minimum 1, etc.). 1) The process called "rotation" has a simple classical meaning; it means transitions from the pocket state Φ_1 to the other pocket state Φ_2 or Φ_3 by means of a reorientation. 2) The second process, called "flip-flop" motion has no classical analogue; it describes the evolution with time of the pocket state after a transition to the first excited librational state. As described in the last paragraph of the preceeding section, a very small splitting of the ground-state multiplet is assumed /7.6/. The name "flip-flop" refers to the two states, between which the nonstationary pocket state oscillates

$$\Phi_1^{LIB}(t) = \frac{1}{\sqrt{3}}\left\{|\Phi_A^{LIB}> + |\Phi_{E^a}^{LIB}> \exp(i\omega_1 t) + |\Phi_{E^b}^{LIB}> \exp(-i\omega_1 t)\right\} . \qquad (7.7)$$

Here ω_1 denotes the tunnel splitting in the excited state. E^a and E^b denote the complex conjugate pair of E states. The evolution is different for the states with E symmetry, which leads to a phase difference between A and E states. The "flip-flop" mechanism therefore is made responsible for the broadening of the A-E transition, yet does not affect E^a-E^b transitions /7.6/. In order to account for the quasielastic scattering observed in MDBP /7.11/, CLOUGH et al. introduced a strain-induced E^a-E^b splitting. "Rotation", which is not necessarily restricted to the classical values for angular steps $\pm 2\pi/3$, does influence both kinds of transition matrix elements.

Flip-flop motion is made responsible for the broadening in Allen's treatment which, however, fails to explain a quasielastic peak at low temperatures /7.6/. Randomly occurring excursions to the excited state lead to a mean additional phase angle denoted by x in (7.6c) and yield a contribution a to the linewidth. In analogy to this and in order to explain the quasielastic scattering a width b in connection with "rotation" is postulated. The same form as in (7.6c) is used. No rigorous derivation is given, though. Initially the same activation energy was introduced for both relaxation processes /6.26,27/. This assumption has been dropped after new experimental evidence /6.28/ of methyl group tunneling in MDBP. The meaning of the two activation energies has been discussed in Sect.6.2.1b. Clough's picture is supported by the fact that the quasielastic linewidth (due only to the second relaxation process) is smaller than that of the inelastic peaks /6.27/.

A further refinement of the model followed the observation of symmetry-dependent lifetimes of the excited librational states /6.27/. Because in Allen's model the lifetime is independent of the spin state of the molecule /6.23/, a modification of the dynamical averaging model appeared necessary. CLOUGH et al. /6.27/ did that by generalizing the relaxation rate introduced in Allen's model [matrix A_2 in (7.3)]. The quantity w_1 which is independent of the spin state is replaced by $w_i+w_i^A+w_i^E$ [with (i = 0,1)]. A derivation analogous to that of /6.23/ is given, based, however, on an equation of motion for the density matrix P. With the same relation between w_0 and w_1 as in Allen's derivation and, additionally, $w_i^A \equiv w_i^E = 0$, Allen's result can be reproduced. The more general model is found to be consistent with the observations in MDBP, taking only $w_i^A = 0$. Details can be found in /6.26,27/.

7.1.3 Coupling to Collective Modes

In the following we shall describe attempts to explain the observed temperature dependence by a coupling to 1) acoustic phonons /6.11/, 2) librational excitations of the crystal /7.1/, and internal vibrations of the molecule /7.2/. The role of transitions to the excited librational states of the molecule under consideration (and of translational optic modes) has been discussed before and will not be repeated here.

HÜLLER /6.11/ emphasized the difference between the polaron problem /7.7/ and tunneling, where symmetry is due to the molecule itself. Remaining in the language of the polaron effect, nevertheless, he distinguished between a "distortion" of the potential (modulates its amplitude) and a "shaking" (modulates the phase ϕ_n in the potential). Only the latter gives rise to a bilinear coupling term in an expansion of the potential $V(\underline{u}_R,\{\underline{u}_i\})$ into the rotational displacement $\underline{u}_R$ and the translational displacements $\underline{u}_i$. The curly brackets signify that the potential V depends on the set of translational displacements of all particles in the crystal.

First the problem of a CH_3 group coupled to a single translational oscillator with $\{\underline{u}_i\} = u_T$ is considered. Then the kinetic energy in the Hamiltonian is

$$K = \frac{1}{2}\,\Theta\dot{u}_R^2 + \frac{1}{2}\,m\dot{u}_T^2 \quad , \tag{7.8}$$

and the harmonic part of V expanded around the minimum in one pocket is

$$V_h(u_R,u_T) = \alpha_1 u_R^2 + \alpha_2 u_T^2 + \alpha_3 u_R u_T \quad . \tag{7.9}$$

Contours of the potential are shown in Fig.7.2; their skewness originates from the coupling term $\alpha_3 u_R u_T$. Pocket state functions are taken to describe the molecule

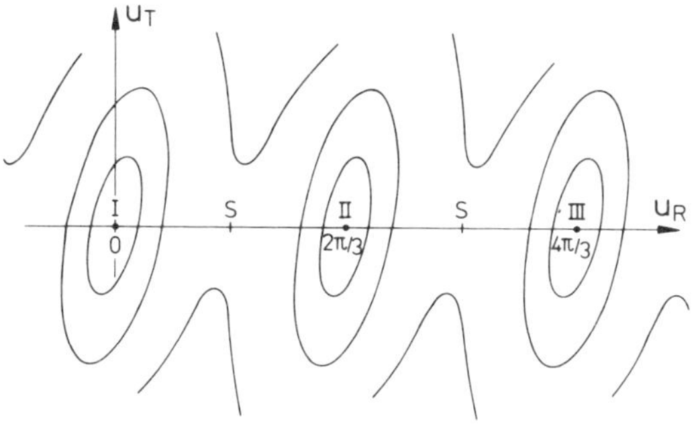

Fig.7.2. Contours of the potential $\Phi(u_R, u_T)$. Minima at $u_T = 0$ and $u_R = 0$, $2\pi/3$ and $4\pi/3$ are denoted by I, II and III; S denotes saddle points. The coupling term $\alpha_3 u_R u_T$ (7.9) is responsible for the skewness of the contours /6.11/

in its rotational ground-state (librational excitations are excluded), while the translational wave function is represented by harmonic oscillator functions specified by a quantum number n_T. States with $n_T > 0$ become increasingly populated with rising temperature. Tunneling frequencies $\omega_0(n_R = 0, n_T)$ are calculated and found to decrease monotonically with n_T. The splitting is reduced by the increasing cancellation of the overlap of the wave functions with increasing n_T; as a function of u_T the sign of Φ changes n_T times in presence of a nonzero coupling.

Second, the Einstein model is replaced by a more realistic Debye model for the translational excitations in the crystal. Only the (translational) displacement u_x of the molecule immersed in a bath of Debye phonons is included. The harmonic potential V_h is rewritten in terms of phonon coordinates /6.11/. Tunneling frequencies then are calculated as a function of phonon occupation. The result for $\hbar\omega_D \gg kT$ (ω_D = Debye frequency) is

$$\omega_p(T) = \omega_0(T=0)\left[1 - \frac{\pi^6}{45}\, g^6\, \frac{\hbar\omega_D}{B}\left(\frac{kT}{\hbar\omega_D}\right)^4\right] . \qquad (7.10)$$

The coupling constant g is a measure for the admixture of translational character to the rotational wave function. It is directly connected with the coupling term between phonon coordinates and angular coordinate. The usually observed decrease of the peak position ω_p of the tunnel line can successfully be explained by a decrease with T^4. On the other hand, the model does not yield a broadening of the states, which is in agreement with results for several NH_4^+-salts, but not for most other systems. The model probably will yield a broadening, too, if a coupling beyond the bilinear term is included. A further generalization dealing with the effect of distortion or the nondiagonal transitions /6.11/ also would be interesting.

A T^4 behavior also has been found by SVARE /7.5/ in a less rigorous treatment. An alternative approach via the coupling to librational excitations of other molecules has been formulated by PUNKKINEN /7.1/. Punkkinen doubted that the rôle of the lifetime w_1^{-1} of the first excited librational state is decisive. Also, ALLEN's model /6.23/ sometimes overestimates the broadening of the tunneling lines. For just two interacting rotors the coupling is represented by a bilinear term $k_{12}u_{R1}u_{R2}$, similar to the case of a coupling to translational modes. Only a weak coupling which yields negligible wave vector dependence of the librational energy is assumed. Therefore an exponential temperature dependence governed by the average librational energy E_1 results. The coupling causes a mixing of the librational states and this gives rise to a weighted average of splittings. No broadening of the levels results, as in the case of a coupling to Debye phonons /6.11/. A width is added later in a very phenomenological way by equating the relaxation rate w_0 with a mean-squared librational amplitude.

The model suffers from several approximations. Also it appears, that in many cases in which tunneling has been observed the dominant contribution to the rotational potential is of a crystal-field type. Contributions depending on the orientations of two molecules often are less important (e.g., ammonium salts). This certainly is not true for van der Waals crystals like methane.

The phenomenological description of the relaxation rate w_0 yields a nonvanishing rate in the limit T = 0, for which zero-point librations are made responsible. This is reminescent of the "adiabatic reorientation processes" postulated by other authors /6.8,7.8,9/. They found a line broadening in the low-temperature limit and claimed a homogeneous broadening due to a finite lifetime $\tau = w_0^{-1}$.

PUNKKINEN has also studied internal vibrations (bending modes) of a molecule in a surrounding of low symmetry /7.2/ and their coupling to rotational tunneling. The coupling is effective only for low site symmetry, which restricts the applicability of the model. Nevertheless it has been demonstrated that the reservoir of modes which may be involved in the mechanism behind the temperature dependence is rather large. At present it is not possible to single out a particular type of excitation as the most important one.

7.2 Pressure Dependence of Tunneling Energies

The pressure dependence of rotational tunneling or, more generally, of all observable rotational transitions of a molecule, provides an interesting access to rotational potentials and intermolecular interactions. As can be seen in Fig.5.9, the tunnel splittings calculated as a function of the potential display an almost exponential behavior. Relatively mild changes of the potential are amplified to rather dramatic changes of the observed tunnel splittings which therefore can be used as a sensitive probe of the rotational potential. More quantitatively this may be phrased in terms of the Grüneisen constant $\gamma = -\partial \ln\omega/\partial \ln V$. Here V denotes the volume of the crystal. For lattice phonons γ is positive and typically $\gamma = 3$. A similar behavior may be expected for librational excitations, whose frequency also increases with decreasing volume. This is different for rotational tunneling in which case the tunnel splitting drops with increasing potential. Using the numbers taken in the estimate below, $\gamma = -20$ ought to be a typical quantity.

For a precise estimate of the shift of the tunneling energies as a function of pressure, one must know the intermolecular forces and the pressure dependence of the structural parameters of the sample. The latter determine the interatomic distances which enter the rotational potential. If the positions and orientations of the atoms and molecules contributing to the rotational potential are not fixed by symmetry, it is necessary to determine these quantities as a function of pressure, too. This would necessitate a sequence of structure analyses accompanying measurements of the tunneling states. If, on the other hand, the structural parameters are fixed by symmetry (which is seldom) or change but little if scaled to the cell parameters, it is sufficient to know the pressure dependence of the cell parameters. For the latter case and cubic symmetry a further simplification arises, and one only needs to know the isothermal compressibility κ.

An estimate based on the knowledge of the compressibility κ and a single power law

$$A_3(a) = A_3(a_0)(a_0/a)^n \tag{7.11}$$

for the strength of the rotational potential has been given by HÜLLER and RAICH /5.14/; a is the lattice parameter at the pressure $p = p_0+\delta p$, where p_0 is a reference pressure. The estimate is given for tetrahedral molecules in a potential $V(\tau) = A_3H_{11}^{(3)}(\tau)$. The general ideas, however, are not restricted to this example. As indicated above, the tunneling frequency ω_0 can be written as

$$\omega_0(A_3) = \omega_0^0(A_3)\exp[g_0(A_3)A_3] \tag{7.12}$$

where $\omega_0^0(A_3)$ and $g_0(A_3)$ depend only weakly on A_3. In particular the product $g_0(A_3)B$ varies little with A_3; it is 0.08 and 0.05 for potentials $A_3 = -25$ B and $A_3 = -125$ B, respectively. For small changes, $\delta A_3/A_3 = -n\delta a/a$ with $\delta a/a = -(1/3)\,\kappa\delta p$, one obtains

$$\delta\omega/\omega = \frac{n}{3}\, g_0(A_3)A_3\kappa\delta p \quad . \tag{7.13}$$

HÜLLER and RAICH gave an estimate for $(NH_4)_2SnCl_6$ /4.27,5.14/, where $A_3 \cong -100$ B, $g_0(A_3) \cong 0.055/B$, $\kappa \cong 0.007$ Kbar^{-1}. A value $\delta\omega/\omega = -5\%/\text{Kbar}$ is found for electrostatic interaction. The leading term in the crystal field experienced by the NH_4^+ group in $(NH_4)_2SnCl_6$ is a monopole-octopole interaction. For an electrostatic origin of this interaction the radial dependence of the crystal field is r^{-n} with $n=4$. A recent measurement of tunnel splittings in $(NH_4)_2SnCl_6$ with neutrons (backscattering) for pressures $p \leqq 2$ Kbar /7.10/ yields an effect twice as strong as expected for purely electrostatic interactions. This indicates important contributions based on overlap or valence forces with a larger value for the exponent n.

Much stronger effects can be predicted for more compressible solids. In methane (CH_4), for example, the compressibility is about 0.04 $Kbar^{-1}$ /6.8/. Pressures of about 5 Kbar would reduce the tunneling energies by almost two orders of magnitude. For this estimate one must not use the linearizations which lead to (7.13). The application of high pressures is less difficult than for solid methane.

The corresponding experiment is difficult and has not yet been performed. Measurements so far have been restricted to pressures of about 1 Kbar and below /6.10/. Such pressures are obtained by pressuring liquid methane up to about 4.0 Kbar and then cooling to low temperatures at constant volume (of the pressure cell). A first result of these measurements, which extend into phase III of CH_4, is the absence of transitions corresponding to almost free rotations above 600 bar. One may conclude that all CH_4 molecules are orientationally ordered in phase III (see also next section).

A beautiful example for the pressure dependence of rotational tunneling has been provided by CLOUGH et al. /7.11/ with sodium acetate trihydrate [$Na(CH_3COO)3H_2O$]. The application of high pressures is less difficult than for solid methane. Measurements with pressures up to 5 Kbar (Fig.7.3) show that the tunneling energies change

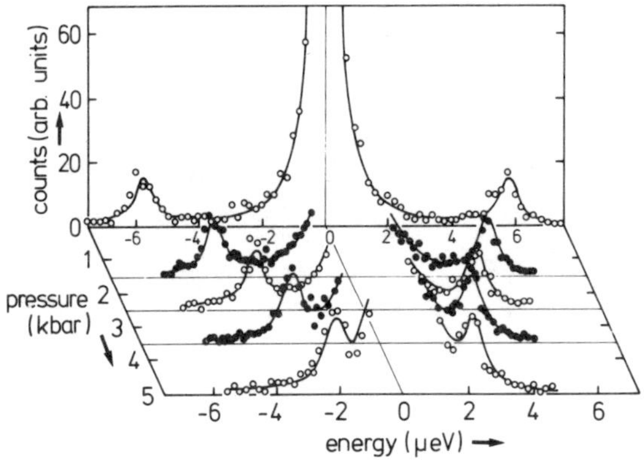

Fig.7.3. Inelastic neutron scattering spectra of sodium acetate trihydrate [$Na(CH_3COO)3H_2O$] at different pressures ($T=4$ K). A pronounced decrease with pressure of the tunneling energy of the methyl group is observed /7.11/

from 5.7 μeV at p = 1 bar to 2.2 μeV at p = 5 Kbar. An almost exponential pressure dependence is found. The authors gave an estimate similar to that of HÜLLER and RAICH /5.14/, this time for a potential $V(\phi) = \frac{1}{2}V_3\cos 3\phi$. The expression

$$\hbar\omega = 316 \exp(-0.116\ V_3) \qquad (7.14)$$

is found to hold approximately in the range studied. In (7.14) $\hbar\omega_0$ is in μeV and V_3 in meV. Assuming an isothermal compressibility $\kappa = 0.01\ \mathrm{Kbar}^{-1}$ as a typical value, a power $n \sim 14$ is obtained. The example probably belongs to the category for which detailed information on the intermolecular interaction can only be safely deduced, if structure analyses are available at least at two different pressures.

The requirement of accurately known interatomic or intermolecular distances represents a serious restriction to the determination of intermolecular interactions. Further experiments are needed for a better judgement concerning the usefulness of the method in this respect. An effort to test the intermolecular interactions in solid nitrogen has recently been published /7.12/. The rotational constant of the nitrogen molecule already is relatively small (B = 0.25 meV), therefore scaled potentials in the ordered phases are large and the classical aspects of solid N_2 dominate. Librational frequencies as a function of pressure and the α-γ transition line are calculated both for a Kihara core potential and for electrostatic quadrupole-quadrupole interaction.

7.3 Isotope Effect

The most direct proof of the presence of tunneling states is the isotope effect /5.16/, due to the characteristically large change of the observed tunnel splittings upon isotopic substitution. An example is the tunneling of $^6Li^+$ and $^7Li^+$ ions in KCl /7.13/; observed energies differ by about 40%. One may ask whether an isotope effect of similar magnitude holds for rotational tunneling.

When discussing the isotope effect, a glance at Fig.5.9 reveals a close relation to the pressure dependence of rotational tunneling, described in the previous section. In both cases relatively moderate changes of the static potential (scaled with $B = \hbar^2/2\Theta$) are amplified to very large changes of the observed splitting. The difference is as follows. By applying pressure the scaled static rotational potential $V'_{st} = V_{st}/B$ is changed in a continuous fashion because V'_{st} depends on the lattice constant. Substitution of the protons in H_2, XH_3 and XH_4 by deuterons, on the other hand, increases the moment of inertia Θ and therefore also the scaled potential

$V'_{st} = 2\Theta V_{st}/\hbar^2$ by a factor of two. Thus deuteration has the same effect as a large change of pressure.

In order to render the effect observable for inelastic neutron scattering, examples with a relatively large splitting for the protonated species must be used. Otherwise the ground-state splitting in the deuterated species is too small to be resolved even with high-resolution neutron spectroscopy. Another possibility is partial deuteration which leads to scaled potentials V'_{st} of intermediate magnitude – but also to a more complex situation, due to the reduced symmetry of the molecule. Recently both neutron and specific-heat experiments have been performed with partially deuterated methane; the former with CH_3D adsorbed on grafoil /6.42/, the other with bulk samples of CH_2D_2 and CHD_3 /7.14-16/.

As an example of the isotope effect we shall discuss deuterated solid methane, which has recently been investigated /5.39,7.17/. A spectrum within the energy range $|E| \leqq 9$ µeV is shown in Fig.7.4. Eight lines at energies 1.20, 2.14, 2.75, 3.37, 4.58, 5.35, 6.70, and 7.95 µeV (statistical error for all lines about ±0.04 µeV) have been obtained from least-squares fits to the data. An extension of the measurements to energy transfers up to 20 µeV did not yield additional lines. A previous measurement /5.39/ covered an energy range $|E| \leqq 4$ µeV and thus only allowed the observation of four transitions. The measuring times are rather long even at a high-flux reactor (about two to three days are required for one spectrum). This is partly due to the complicated spectrum and partly due to the small incoherent scattering cross section $\sigma_{inc} = 2.2$ barn of deuterium.

At this point of the discussion it appears useful to report a theoretical prediction for the isotope effect in methane /5.9,7.10/. The observed splitting in CH_4 II (143 µeV and 73 µeV) can be explained with a rotational potential $V(\tau) = A_3H^{(3)}(\tau)$ with $A_3/B = -37$. Naively, one would expect a doubling of A_3/B in CD_4 because of the change in B, but due to the reduced librational zero point motion, the

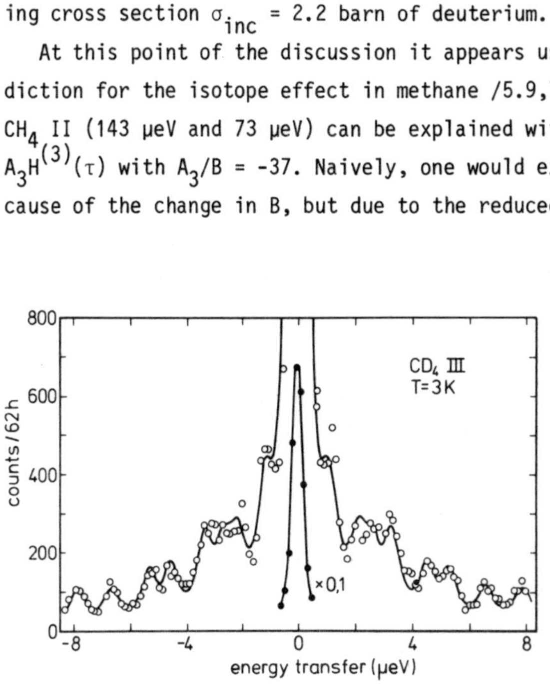

Fig.7.4. Inelastic neutron scattering spectrum of solid CD_4 III, measured with a backscattering spectrometer. The ground-state splitting of the deuterated compound (CD_4) decreases by a factor 30 with respect to the protonated species (CH_4). This decrease reveals a pronounced isotope effect /5.39,7.17/

molecular field in CD_4 is $A_3/B(CD_4) = -85$, more than twice the value in CH_4 /5.14/. The prediction for the tunnel splitting in CD_4 is $\hbar\omega_{TE} = 0.0038 \cdot B = 1.25$ μeV; this shift relative to CH_4 corresponds to an isotope effect of about a factor of 50. As can be seen from the tunnel splittings in CD_4 (T = 3 K), measured by PRAGER et al. /7.17/, the observed isotope effect is about 30, which is in reasonable agreement with the theoretical predictions /5.14/. In any case the isotope effect is very large, much larger than for the ring tunneling of Li^+ ions in KCl /7.13/, for example.

It is a complication that the prediction is based on phase II of methane (Fig. 6.1), while the measurements actually have been performed in CD_4 III. As already indicated, the two structures are believed to be closely related, the main difference being that all molecules are ordered in phase III /2.12,6.10,7.18,19/. It is not clear whether this statement can be maintained in view of the complicated structure of the observed spectra. So far, no simple assignment (e.g., on the basis on MAKI's predicted structure of phase III /7.19/) has been found. Within the model of MAKI et al., large T state splittings appear to be very unlikely as the low-symmetry terms in the potential are very weak.

Instead of fitting peak positions and peak intensities separately, the data analysis can also be based on a fit of the 120° overlap matrix elements h_i (see also Sect.6.2.4b). Different sets of h_i have to be attributed to sublattices which are not symmetry related. A set of h_i determines both the level scheme and – via the transition matrix elements, which recently have been calculated for CD_4 /3.4/ (see Sect.5.3 and the appendix) – the intensity of the transitions. No unambiguous result is obtained. Several models with 3 and 4 inequivalent sublattices fit the data equally well /7.17/. A final answer requires either a structure analysis in phase III of methane or additional high-resolution measurements in CH_4 III.

Three common features of the models appear noteworthy. 1) The specific heats derived from the models are practically indistinguishable. Hence one may not hope for detailed results from specific-heat measurements in systems like CD_4 III (with several inequivalent sublattices). 2) The peaks at large energies always combine into one level scheme with low site symmetry. Because of the large splitting (indicating a weaker potential), a correlation with the sites of disordered molecules in phase II is suggestive. Then, however, one also would expect a relatively high site symmetry. 3) All models generate more than eight transitions, some of which are very close to each other and appear as single lines.

7.4 Tunneling in Molecular Mixtures

We have already discussed rotational excitations of molecules, matrix isolated in atomic crystals (Sect.6.1.3). Similarly one may start from the opposite side, that is, molecular crystals, and ask for the effect of substitutional impurities on low-temperature rotational excitations. It has been mentioned before that paramagnetic impurities, such as oxygen or free radicals created by γ irradiation, couple to the spin functions of the molecules and can speed up nuclear-spin conversion considerably. Here we only want to discuss the effect of nonmagnetic atomic impurities.

The statistical replacement of molecules (which have an anisotropic density distribution) by atoms or monatomic ions (which are isotropic) has several interesting aspects. In analogy to magnetic systems one may expect that the impurities behave like nonmagnetic atoms, destabilize orientationally ordered phases and eventually give rise to phases with spin glass character. Interesting systems of this kind are mixtures of o-H_2 and p-H_2 /7.20/ (where p-H_2 is the isotropic component), $K(CN)_{1-x}Br_x$ /7.21/ and also $(N_2)_{1-x}Ar_x$ /7.22/ and $(CD_4)_{1-x}Kr_x$. Much remains to be done in this field, which is not a topic of the present review, however,

Another interesting aspect is the effect of impurities on the rotational excitations in a molecular crystal. Judging from the effect of pressure, one may suspect that the rotational energies are rather sensitive to impurities. Atoms or monatomic ions which statistically replace molecules give rise to a distribution function of rotational potentials. In general, the influence of substitutional impurities on the crystalline field $V_c(\omega_i^E)$ will be rather weak. This is due to the fact, that good mixing occurs only if the substituted and the substituting particles have the same charge and if their ionic or van der Waals radii are rather similar, too. A larger effect will concern interactions of the type $V(\omega_i^E,\omega_j^E)$ (see Sect.2.3) to which atoms or monatomic ions do not contribute. Therefore the summation in

$$V^i(\omega_i^E) = \sum_j V(\omega_i^E,\omega_i^E)$$

is over fewer neighbors than in the unperturbed crystal.

This gives rise to a distribution function of rotational potentials and thus also to a spectrum of energy eigenvalues, that is, to a (inhomogenous) broadening of the lines which depends on the impurity concentration.

Ideally the substitution only affects the orientation-dependent interaction and no position-dependent interactions. This is almost perfectly fulfilled in mixtures of o-H_2 and p-H_2 /7.20/ and $(NH_4)_2SnCl_6$ and K_2SnCl_6 /7.23/ where the lattice constants of the two components are very similar. For most other substances a local lattice relaxation around the impurity will result.

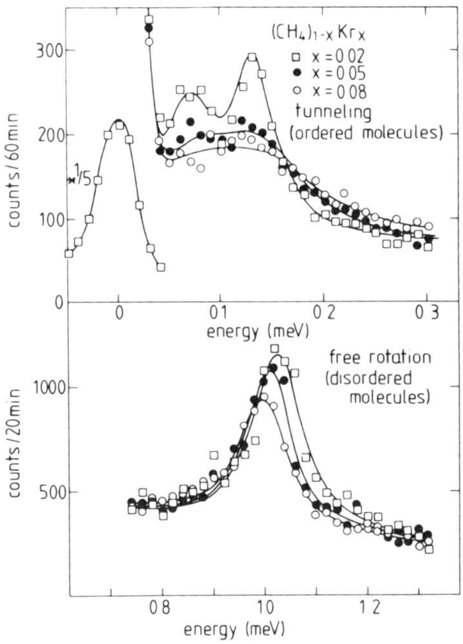

Fig.7.5. Effect of krypton impurities on the tunneling lines and the rotational excitations in CH_4 II /7.24/

The effect of impurities in a system with dominant multipole-multipole interactions is shown in Fig.7.5. The figure displays the low-lying rotational excitations in the mixed system $(CH_4)_{1-x}Kr_x$ with $x = 0.02$, 0.05, and 0.08 /7.24/: 1) the tunneling transitions of the ordered molecules and 2) the perturbed 0-1 rotational transition of the disordered molecules. From specific-heat measurements /7.25/ it is known that phase II is destabilized for $x \geqq 0.15$ /7.23/. Thus for the reported experiments the binary mixture is in phase II. One may note that both tunneling states and almost free rotation broaden with increasing concentration [e.g., for tunneling: $\Gamma(x = 0.02) = 26 \pm 4$ μeV, $\Gamma(x = 0.05) = 50 \pm 6$ μeV, $\Gamma(x = 0.08) = 63 \pm 15$ μeV]. The tunneling lines shift to larger energies which can be explained by a weaker average potential in the presence of krypton impurities. For the 0-1 rotational transition, on the other hand, a reduction of the energy with impurity concentration results. This indicates a stronger potential and shows that the cancellation of the octopole-octopole interaction at the sites of disordered molecules does not work as perfectly in the binary mixture as in pure methane (see also Sects.6.1.2 and 6.2.3a).

There is not only an effect on the average magnitude of the rotational potential, but also on its symmetry. Statistical replacement of molecules leads to a binomial distribution of the number of neighbors of a given kind. On average the surroundings of a molecule retain high symmetry, but locally the symmetry will be perturbed. For tetrahedral molecules this gives rise to a T state splitting, which adds to the broadening of A-T at T-E transitions (after averaging over many configurations), but also gives rise to inelastic transitions within the T state multiplet. The calculation of rotational spectra for binary mixtures requires the calculation of energy eigenvalues for many configurations of low symmetry with a subsequent averaging over the configurations (inhomogenous broadening). Such calculations have not been performed yet.

For large impurity concentrations ($x \cong 0.5$) one may expect a broad distribution of potentials and thus very broad spectra. If one may ignore the crystal field, and attributes the main contribution to the potential to anisotropic interactions between molecules, the neutron spectra will look almost like quasielastic scattering. This indeed has been found for $(N_2)_{0.7}Ar_{0.3}$ /7.26/.

Here an interesting aspect appears. Quasielastic spectra at low temperatures imply an almost constant density of rotational states at low energies and consequently one should observe an almost linear specific heat for $T \ll \Theta_D$ (Θ_D = Debye temperature). This means that one can start with a molecular crystal with low-energy rotational excitations (e.g., tunneling states in CH_4) which give rise to Schottky anomalies, and by admixing impurities one may obtain a solid with a linear specific heat at low temperatures. Apparently there is a similarity to glasses, in which the observed linear specific heat is attributed to the presence of a distribution function of two-level systems (tunneling states /7.27/). These low-energy states, however, are absent in the respective crystalline system (e.g., of SiO_2) and no continuous transition from the crystalline to the glassy state is possible. In a crystal like SiO_2, the only low-lying energy states are associated with acoustic phonons, which gives rise to the usual T^3 behavior of the specific heat.

Another binary system which has recently been studied /7.23/ is $(NH_4)_{2-2x}K_{2x}SnCl_6$. As becomes obvious from Fig.7.6, the impurities have only little effect on the tunneling spectrum. Figure 7.7 shows both the tunnel splitting $\hbar\omega_{T-E}$ and the width of the lines for concentrations $x \leq 0.6$. The findings indicate that the octopole-octopole interaction between two NH_4^+ tetrahedra in $(NH_4)_2SnCl_6$ (Sects.4.5.1 and 6.2.4a) provides only a minor contribution to the potential and that the rotational potential largely has crystal-field character. This is due to the strong monopole-octopole interaction which does not change when a NH_4^+ group is replaced by an isoelectronic potassium ion.

At concentrations $x \cong 0.65$ a phase transition to the low-temperature phase of K_2SnCl_6 /7.28/ takes place, the structure of which is not yet completely established. The resulting distortions seem to cause an increase of the rotational potential and, consequently, a strong decrease of the tunneling energies. As an upper limit a value $\hbar\omega_{A-T} = 0.3\ \mu eV$ may be given. In principle one could use the NH_4^+ ion as a probe for the site symmetry at the potassium or rubidium sites. In the present case (K_2SnCl_6), however, the energy resolution available in a neutron backscattering experiment has not been sufficient.

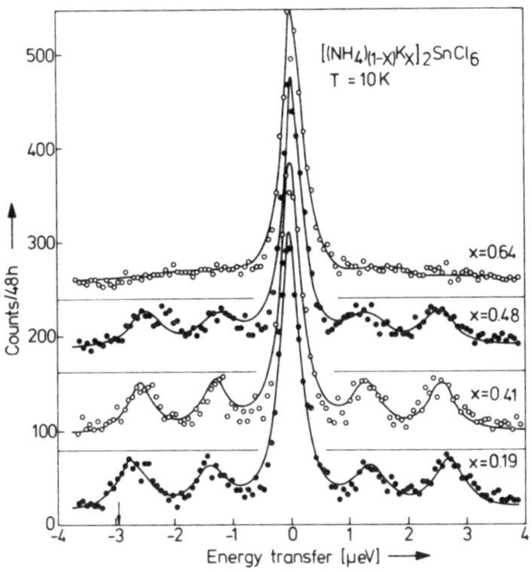

Fig.7.6. Tunneling spectra of the NH_4^+ group in the mixed crystal $(NH_4)_{2-2x}K_{2x}SnCl_6$ for several concentrations x /7.23/ (temperature T = 10 K, momentum transfer Q = 1.9 Å^{-1})

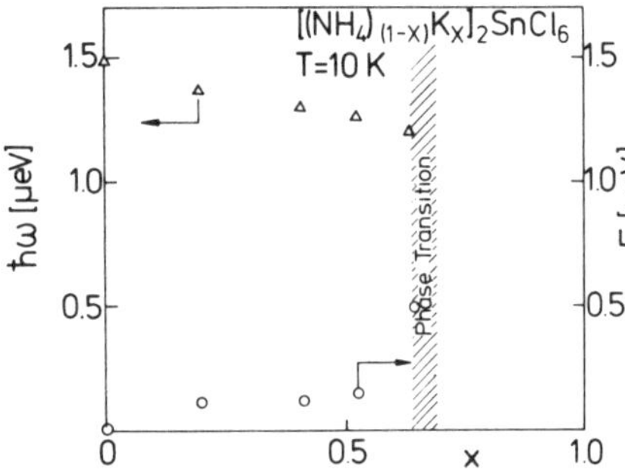

Fig.7.7. Variation of the ground-state splitting $\hbar\omega_{T-E}$ and of the linewidth Γ (HWHM) as a function of the potassium concentration x /7.23/. See Fig.7.6

Appendix: Calculation of Transition Matrix Elements

A1. Calculation of Spin Functions for XH_4 and XD_4

The spin functions for tetrahedral molecules can be constructed as follows (a similar calculation applies to other molecules with internal symmetries). First we note that there are four different types of spin states:

$$\chi_1 = |aaaa\rangle \tag{A.1a}$$

$$\chi_2 = |baaa\rangle \tag{A.1b}$$

$$\chi_3 = |bbaa\rangle \tag{A.1c}$$

$$\chi_4 = |bbac\rangle \ . \tag{A.1d}$$

For protonated molecules a and b denote either of the eigenvalues 1/2 or -1/2 of the z component of the proton spins. Only the first three types of spin states χ_i exist (total number $2^4 = 16$). For deuterated molecules a, b and c denote one of the eigenvalues +1, 0, -1. All four types χ_i exist (total number $3^4 = 81$). Now symmetrized functions are obtained by use of the projection operator $P^\Gamma \sim \sum_{\mu=1}^{12} \alpha_{ij}^{\Gamma*}(\mu)\hat{R}_\mu$. Here the sum is over all 12 symmetry operations $\hat{R}_\mu$ of the tetrahedral group; Γ denotes the representations A, E and T, respectively, and $\alpha_{ij}^{\Gamma}(\mu)$ the elements of the representation matrices. For one-dimensional representations the α_{ij}^{Γ} are the characters of the representation.

The functions obtained by the use of P^Γ are eigenfunctions of the nuclear-spin operator $\mathring{i}_z$, but not necessarily of $\mathring{i}^2$ as well. Obviously $|aaaa\rangle$ is completely symmetric and has A symmetry. The functions of the type $|baaa\rangle + |abaa\rangle + |aaba\rangle + |aaab\rangle$ are also completely symmetric. Similarly, the functions $|baaa\rangle + |aaab\rangle - |aaba\rangle - |abaa\rangle$ have symmetry T_x. From χ_1 and χ_2 no spin functions of E symmetry can be constructed. The above functions are also eigenfunctions of $\mathring{i}^2$ in the case of CH_4, and for $|I_z| \geq 2$ also for CD_4. The construction of spin functions $|I,I_z,\Gamma\rangle$ is more difficult for $I_z = 0$ (CH_4 and CD_4) and $I_z = \pm 1$ (CD_4). "Families" of spin functions with total spin I can be generated by repeated application of the totally symmetric lowering operator $\mathring{i}_- = \sum_{\gamma=1}^{4} i_{\gamma-}$ and use of orthogonality. This method is needed for CD_4, where 81 symmetrized spin functions have to be constructed: 15 A functions with I = 4,2 and 0, 3×18 T functions with I = 3, 2, 1 and 1, and 2×6 E functions with I = 2 and 0. Only the spin functions for CH_4 are listed in this work (Table 5.6). If one is only interested in the total intensity for a transition between states of different symmetry — one step, namely, the construction of functions which are also eigenfunctions of $\mathring{i}^2$, can be omitted. Wave functions ψ, which are completely symmetric under the symmetry operations of the tetrahedral group, are obtained as products of spin functions and rotational wave functions of the same symmetry $\psi = \Phi^{\Gamma} \cdot \chi^{\Gamma}$. As in /3.3/ δ functions have been used for the spatial pocket states, which is inadequate for the calculation of overlaps, but often represents a good approximation on the way to transition matrix elements.

In practical calculations of transition matrix elements a somewhat modified approach has been used. In the spin functions defined above, μ_i denotes the z component of the ith proton. Instead, functions $[\mu_1\mu_2\mu_3\mu_4>$ are used /3.3/, where μ_i denotes the z component of the ith site and not that of an individual proton (Sect.5.2). With this definition the spatial part of the wave function $|\Phi>$ is simply included as a function sharply peaked at the equilibrium orientation τ_0 of the tetrahedron. For complete orientational localization $|\Phi> \sim \delta(\tau-\tau_0)$, independent of the symmetry of the rotational state considered. This modified approach takes into account that ultimately only the spin state at a given site determines the matrix elements and leads to a considerable simplification of the calculations. A generalization to finite size wave functions $|\Phi>$ is demonstrated in /3.4/.

A2. Calculation of Transition Matrix Elements (Cubic Symmetry)

The next step consists in the application of the neutron scattering operator $W = W_A + W_T$ (Sect.5.3) to the wave functions $\psi = |\Phi>[\chi_\Gamma(I,I_z)>$. W_A does not change the symmetry of the spin functions $\chi_\Gamma(I,I_z)$, while it is always changed by W_T. Matrix elements are obtained by decomposing the resultant functions $(W_A+W_T)\psi$ into symmetry adapted functions $\psi' = |\Phi>[\chi_{\Gamma'}(I',I_z')>$. In practice this calculation is done by means of a computer program. In particular the 81×81 matrix of transition elements for CD_4 necessitates such an approach. The resultant 16×16 matrix $\bar{\bar{A}}$ for CH_4 is explicitly given in Table A.1. For reasons of spin conservation the scattering obeys the selection rules $\Delta I = 0, \pm 1$ and $\Delta I_z = 0, \pm 1$. Transitions caused by $\mathring{s}_z \cdot \mathring{i}_z$ (Sect.5.3) leave both z-component μ of the neutron spin and I_z unchanged (scattering without spin flip). However, the "non-spin-flip" part of W_T may change the total spin I of the molecule.

Table A.1. Transition matrix elements for CH_4. Initial and final states are described by their symmetry Γ and Γ', respectively, and by the z component of the nuclear spin I_z and I_z', respectively; $a = 1$, $b = 1/2$, $c = 1/\sqrt{3}$, $d = 1/\sqrt{6}$, $e = 1/\sqrt{2}$, $f = \sqrt{4/3}$, $g = \sqrt{3/2}$, $h = \sqrt{2/3}$. For transitions without symmetry change the elements have to be multiplied by G_A, otherwise by G_{T_x}, G_{T_y} and G_{T_z} (indices x, y and z denote which one has to be taken). When calculating intensities, it should be noted that transitions without spin-flip occur both for neutrons with spin up $|\alpha\rangle$ and spin down $|\beta\rangle$. Matrix elements which are zero are omitted from the Table. The corresponding 81×81 matrix for CD_4 is not listed explicitly (its T-T part is given in /3.4/). Intensities for powder samples can be found in Table A.2

Γ	Γ' / I_z' — I_z	A					T_x			T_y			T_z			E	
		-2	-1	0	1	2	-1	0	1	-1	0	1	-1	0	1	0	0
A	-2	-a	a				a_x			a_y			a_z				
	-1	a	-b	g			b_x	e_x		b_y	e_y		b_z	e_z			
	0		g		g		$-d_x$	c_x	d_x	$-d_y$	c_y	d_y	$-d_z$	c_z	d_z		
	1			g	b	a		$-e_x$	b_x		$-e_y$	b_y		$-e_z$	b_z		
	2				a	a			$-a_x$			$-a_y$			$-a_z$		
T_x	-1	$-a_x$	b_x	$-d_x$			-b	e		b_z	$-e_z$		b_y	$-e_y$		$-c_x$	a_x
	0		e_x	c_x	$-e_x$		e		e	$-e_z$		$-e_z$	$-e_y$		$-e_y$	$-d_x$	c_x
	1			d_x	b_x	$-a_x$		e	b		$-e_z$	$-b_z$		$-e_y$	$-b_y$	c_x	$-a_x$
T_y	-1	a_y	b_y	$-d_y$			b_z	$-e_z$		-b	e		b_x	$-e_x$		$-c_y$	$-a_y$
	0		e_y	c_y	$-e_y$		$-e_z$		$-e_z$	e		e	$-e_x$		$-e_x$	$-d_y$	$-e_y$
	1			d_y	b_y	$-a_y$		$-e_z$	$-b_z$		e	b		$-e_x$	$-b_x$	c_y	a_y
T_z	-1	a_z	b_z	$-d_z$			b_y	$-e_y$		b_x	$-e_x$		-b	e		f_z	
	0		e_z	c_z	$-e_z$		$-e_y$		$-e_y$	$-e_x$		$-e_x$	e		e	h_z	
	1			d_z	b_z	$-a_z$		$-e_y$	$-b_y$		$-e_x$	$-b_x$		e	b	f_z	
E	0						$-c_x$	$-d_x$	c_x	$-c_y$	$-d_y$	c_y	f_z	h_z	f_z		
	0						a_x	c_x	$-a_x$	$-a_y$	$-e_y$	a_y					

Transitions caused by $\overset{\circ}{s}_+\overset{\circ}{i}_{\gamma-}$ and $\overset{\circ}{s}_-\overset{\circ}{i}_{\gamma+}$ are associated with a change of both μ and I_z (spin-flip scattering). The diagonal blocks in the matrix are mediated by the totally symmetric part of the operator W_A. They all are proportional to $G_A = \frac{1}{2}\sum_\gamma G_\gamma$ with $G_\gamma = \exp(i\underline{Q}\cdot\underline{r}_\gamma)$ and $\underline{r}_\gamma$ denoting the positions of the four protons of a tetrahedron.

The off-diagonal blocks are obtained by application of the operators W_{T_x}, W_{T_y} and W_{T_z} (see Sect.5.3). Action of W_{T_x} on a function $[\chi_{T_y}\rangle$ leads to functions $[\chi_{T_z}\rangle$ and a factor $G_{T_x} = \frac{1}{2}(G_1-G_2-G_3+G_4)$. The other blocks can be constructed by cyclic permutations or have to be transposed for noncyclic permutations. Finally, intensities are obtained by taking the modulus squared of the matrix elements and then summing within a block connecting the symmetries Γ and Γ'. This summation yields symmetric matrices of dimension 5 which are given for both CH_4 (Table A.2) and CD_4 (Table A.3). The $\underline{Q}$ dependence of the scattering is contained in expressions $|G_\Gamma|^2 = 1+0.5\sum_{i\neq j}\sigma_i^\Gamma\sigma_j^\Gamma \cos \underline{Q}\cdot\underline{r}_{ij}$, with $\underline{r}_{ij} = \underline{r}_i-\underline{r}_j$; the σ_i^Γ are ±1, with the pattern of signs defined in (5.8).

Table A.2a,b. Intensities for transitions between the group of states with symmetry Γ and Γ' of a tetrahedral molecule; a common factor $\sigma_{inc}/216$ has been extracted from both tables for (a) CH_4 and (b) CD_4. The following abbreviations have been used: $g_A^2 = |G_A|^2$, $g_x^2 = |G_{T_x}|^2$, $g_y^2 = |G_{T_y}|^2$ and $g_z^2 = |G_{T_z}|^2$. The table in its present form assumes $T \gg \Delta$ (Δ is a measure for the tunnel splitting) in which case the population of all ground-state levels is the same. A generalization to lower temperatures with proper inclusion of Boltzmann factors is straightforward

(a) CH_4	A	T_x	T_y	T_z	E
A	$135g_A^2$	$45g_x^2$	$45g_y^2$	$45g_z^2$	0
T_x	$45g_x^2$	$27g_A^2$	$27g_z^2$	$27g_y^2$	$36g_x^2$
T_y	$45g_y^2$	$27g_z^2$	$27g_A^2$	$27g_x^2$	$36g_y^2$
T_z	$45g_z^2$	$27g_y^2$	$27g_x^2$	$27g_A^2$	$36g_z^2$
E	0	$36g_x^2$	$36g_y^2$	$36g_z^2$	0

(b) CD_4	A	T_x	T_y	T_z	0
A	$70g_A^2$	$30g_x^2$	$30g_y^2$	$30g_z^2$	0
T_x	$30g_x^2$	$42g_A^2$	$42g_z^2$	$42g_y^2$	$36g_x^2$
T_y	$30g_y^2$	$42g_z^2$	$42g_A^2$	$42g_x^2$	$36g_y^2$
T_z	$30g_z^2$	$42g_y^2$	$42g_x^2$	$42g_A^2$	$36g_z^2$
E	0	$36g_x^2$	$36g_y^2$	$36g_z^2$	$20g_A^2$

As most measurements use powder samples, it is useful to quote powder averaged quantities as well. Defining $g_\Gamma^2 = |G_\Gamma|^2$, and $\overline{g_\Gamma^2}$ as its powder average one obtains

$$\overline{g_A^2} = 1 + 3\,j_0\,(Q\rho\sqrt{8}/\sqrt{3}) \tag{A.2}$$

$$\overline{g_T^2} \equiv g_{T_x}^2 + g_{T_y}^2 + g_{T_z}^2 = 1 - j_0\,(Q\rho\sqrt{8}/\sqrt{3}) \quad . \tag{A.3}$$

Table A.3a,b. Intensity of T-T transitions for CH_4 (and CD_4) for reduced site symmetries. (a) refers to a threefold symmetry axis; T_1 denotes the nondegenerate level, T_2 and T_3 the degenerate ones. (b) refers to a twofold symmetry axis; T_1, T_2 and T_3 denote the states with energy eigenvalues $-\delta$, 0 and $+\delta$, respectively. A common factor $\sigma_{inc}/216$ has been extracted from both tables. The corresponding tables for CD_4 are obtained by multiplication with 14/9. The intensities of A-T_i and E-T_i transitions (Table A.2) remain unaffected by reduced symmetry

Γ \ Γ'	T_1	T_2	T_3
T_1	$63+45\ j_0$	$9-9\ j_0$	$9-9\ j_0$
T_2	$9-9\ j_0$	$54+54\ j_0$	$18-18\ j_0$
T_3	$9-9\ j_0$	$18-18\ j_0$	$54-54\ j_0$

Γ \ Γ'	T_1	T_2	T_3
T_1	$54+54\ j_0$	$27-27\ j_0$	0
T_2	$27-27\ j_0$	$27+81\ j_0$	$27-27\ j_0$
T_3	0	$27-27\ j_0$	$54+54\ j_0$

Here j_0 denotes a spherical Bessel function and ρ is the radius of a molecule. For cubic symmetry, all T states are degenerate and therefore all T-T transitions are elastic. Only the part resulting from the totally symmetric operator W_A corresponds to the elastic intensities in the classical limit. The same statement holds also for CH_3 groups. One may suspect that this relation is important in connection with models for the temperature dependence of rotational tunneling.

A3. Transitions at Sites with Reduced Symmetry

So far, only the intensity for cubic site symmetry has been calculated. As we shall see somewhat later, the availability of intensity information in addition to the positions of the inelastic transitions is important for reduced site symmetries. In the following, a simple method for the calculation of intensities is given. It starts from the transition matrix elements in the cubic case. Only the T states are affected by lower symmetry. Degeneracies are removed, and as a consequence some or all of the transitions between the T states (with energy E_{T_i}) become inelastic. The spatial wave functions Φ_{CUB} combine to new states Φ_N

$$\Phi_N = \bar{\bar{R}}^N \Phi_{CUB} \quad . \tag{A.4}$$

The matrix $\bar{\bar{R}}^N$ diagonalizes the Hamiltonian matrix $\hat{\mathscr{H}}_T$ (Table 5.3) and thus can easily be obtained.

Similarly, the spin function χ_{CUB} (both $[\chi>$ and $|\chi>$) combine to new states χ_N

$$\chi_N = \mathcal{R}\,\chi_{CUB} \tag{A.5}$$

with

$$\mathcal{R} = \begin{pmatrix} \bar{\bar{R}}_{11} & \bar{\bar{R}}_{21} & \bar{\bar{R}}_{31} \\ \bar{\bar{R}}_{12} & \bar{\bar{R}}_{22} & \bar{\bar{R}}_{32} \\ \bar{\bar{R}}_{13} & \bar{\bar{R}}_{23} & \bar{\bar{R}}_{33} \end{pmatrix} \tag{A.6}$$

Here $\bar{\bar{R}}_{ij} = \bar{\bar{R}}^N_{ij}\bar{\bar{E}}$; $\bar{\bar{E}}$ is the unit matrix of dimension 3 for CH_4 and of dimension 18 for CD_4, respectively. Transition matrix elements which are adapted to the given symmetry are obtained by rotating the T-T part $\bar{\bar{M}}$ of the relevant matrix $\bar{\bar{A}}$

$$\bar{\bar{M}}_N = \mathcal{R}^T\bar{\bar{M}}\mathcal{R} \quad . \tag{A.7}$$

For CH_4, this matrix is given in Table A.1; the corresponding 54×54 matrix for T-T transitions in CD_4 is not listed explicitly.

Only for two cases will the results be presented in the form of tables and in both cases the 180° overlap matrix elements H are neglected: a) for a threefold axis at the crystal site which yields $h_1 \neq h_2 = h_3 = h_4$; b) for a twofold axis at the crystal site which causes the relation $h_1 = h_2 \neq h_3 = h_4$ between the 120° overlap matrix elements h_i. For more than two independent elements h_i, the matrix $\bar{\bar{R}}_N$ depends not only on the symmetry but also on the relative magnitude of the overlap matrix elements h_i. For simplicity only powder-averaged quantities are given in Table A.3. One may note that—apart from a factor—the results are the same for XH_4 and XD_4.

Relations between the 120° overlap matrix elements, which contain information on the site symmetry, cannot always be concluded unambiguously from the energy-level scheme. A "symmetric" T state splitting does not necessarily mean the presence of a twofold axis. For an unambiguous assignment, the intensity of the observed transitions has to be analyzed in addition to the level scheme. If there is a twofold axis at the molecular site, one T-T transition is forbidden (Table A.3). Unambiguous conclusions (on the basis of peak positions only) are possible if two or all three T states are degenerate.

The use of intensity information is even more important if the 180° overlap matrix elements H_i cannot be neglected and additional parameters enter. Sizeable contributions from H, however, signal a relatively large width of the pocket states and inclusion of this width in the calculations may be required.

A4. Transition Matrix Elements for the Neutron Scattering from Methyl Groups

In a calculation of the neutron scattering from tunneling CH_3 groups or NH_3 molecules one proceeds as in Sect.A.1. Only the results of a calculation with δ type pocket states shall be quoted. The spin functions may be decomposed into states with A, E^a and E^b symmetry which are listed in Table 5.6. Similarly the neutron scattering operator may be written $W = W_A + W_E$, with

$$W_A = \frac{1}{3}\,\mathring{s}(\mathring{i}_1+\mathring{i}_2+\mathring{i}_3)(G_1+G_2+G_3) \tag{A.8}$$

$$W_{E^a} = \frac{1}{3}\,\mathring{s}(\mathring{i}_1+\varepsilon\mathring{i}_2+\varepsilon^*\mathring{i}_3)(G_1+\varepsilon^*G_2+\varepsilon G_3) \tag{A.9}$$

W_{E^b} is the complex conjugate of W_{E^a} and $G_\gamma = \exp(i\underline{Q}\cdot\underline{i}_\gamma)$. In a way analogous to Sect.A.1 we define

$$|G_A|^2 = 1 + \frac{1}{3}\sum_{i\neq j} \cos\underline{Q}\cdot\underline{r}_{ij} \tag{A.10}$$

$$|G_{E^b}^{\ a}|^2 = 1 + \frac{1}{3}\sum_{i\neq j} \cos(\underline{Q}\cdot\underline{r}_{ij} \pm 2\pi/3) \tag{A.11}$$

with $\underline{r}_{ij} = \underline{r}_i - \underline{r}_j$. The corresponding powder-averaged quantities are

$$\overline{g_A^2} = 1 + 2\,j_0(Q\rho\sqrt{3}) \tag{A.12}$$

$$\overline{g_E^2} = \overline{g_{E^a}^2} + \overline{g_{E^b}^2} = 2 - 2\,j_0(Q\rho\sqrt{3}) \quad . \tag{A.13}$$

The intensities for the possible transitions are given in Table A.4. For a powder sample, the scattering function is

$$S_{inc}(\underline{Q},\omega) = \left[\frac{5}{3}+\frac{4}{3}j_0(Q\rho\sqrt{3})\right]\delta(\omega) + \left[\frac{2}{3}-\frac{2}{3}j_0(Q\rho\sqrt{3})\right]\left[\delta(\omega-\Delta)+\delta(\omega-\Delta)\right] \quad . \tag{A.14}$$

Here Δ denotes the tunnel splitting. In this expression an equal population of all ground-state levels is assumed which is fulfilled for $kT \gg \Delta$. On the other hand, the temperature has to be sufficiently low to allow the neglect of fluctuations of the rotational potential.

Table A.4. Intensity of the transitions between the different symmetry states Γ and Γ' of a CH_3 group or a NH_3 molecule. Intensities for powder samples are obtained by averaging the quantities $G_{\Gamma}^2(\underset{\sim}{Q})$ over all orientations of $\underset{\sim}{Q}$. The values given in the table refer to $T \gg \Delta$ (Δ = tunnel splitting) and thus are based on equally populated ground-state levels

Γ \ Γ'	A	E_a	E_b
A	$\frac{5}{2}\, G_A^2$	$G_{E^b}^2$	$G_{E^a}^2$
E_a	$G_{E^b}^2$	$G_{E^b}^2$	$\frac{1}{4}\, G_A^2$
E_b	$G_{E^a}^2$	$\frac{1}{4}\, G_A^2$	$G_{E^a}^2$

List of Symbols

a_{coh}	Coherent scattering length
a_{inc}	Incoherent scattering length (rather a_{sd} = spin-dependent scattering length)
a_+, a_-	Scattering length for parallel and antiparallel orientation of neutron and nuclear spin
$A^{n\gamma}$	Spin-dependent nuclear interaction (neutron-nucleus)
$\overline{A}$	Averaged nuclear interaction
$A(\underline{Q})$	Structure factor
α	Angle, angular step in diffusion process
$\bar{\bar{A}}, \bar{\bar{A}}_1, \bar{\bar{A}}_2$	Matrices containing frequencies and relaxation rates of tunneling CH_3 group
B	Rotational constant
ce_i	Mathieu functions (even)
D_R	Rotational diffusion constant
D_T	Translational diffusion constant
d	Dimension of unit sphere (e.g., d = 2 is a unit circle)
Δ	Splitting of two-level system *and* energy of T-E transition for tetrahedral molecules

δ_p	Shift of peak position for tunneling transition
$\mathcal{D}^{(\ell)}_{mm'}(\omega^E)$	Wigner functions
$\frac{d^2\sigma}{d\Omega d\varepsilon}$	Double differentials scattering, cross-sections (d = distinct, s = self)
$d\Omega$	Solid angle element
$d\varepsilon$	Energy interval
E	Energy transfer of the neutrons
E^A	Activation energy
E_i	Energy of ith librational state
$\overline{E}_i$	Mean energy of ith librational state
$E_{J_1J_2}$	Transitions of almost freely rotating molecule
E_{AT}, E_{TE}	Transitions within ground-state multiplet of tetrahedral molecule
E_A, E_E, E_T	Ground-state energy of tetrahedral molecule
ΔE	Energy resolution
EQQ	Electrostatic quadrupole-quadrupole interaction
EOO	Electrostatic octopole-octopole interaction
EISF	Elastic incoherent structure factor

$\exp(-2w)$	Debye-Waller factor
η	Order parameter
$F^{\ell}_{m'm}(t), F_{\ell}(t)$	Rotational relaxation functions
$f(\omega^E)$	Distribution function of orientations ω^E
$\underline{F}(t)$	Fluctuating force field
$\underline{f}_0(\omega^E)$	Static force field
$F(\underline{Q})$	Form factor
ϕ	Rotational angle
ϕ_n	Phase angle
Φ, Φ_{ℓ}	Rotational wave function
$G_S(\underline{r},t)$	Van Hove self-correlation function
$G_S(\underline{r},\underline{r}_0;t)$, $G_S(\omega^E,\omega^E_0;t)$	Generalized self-correlation function
$g(\underline{r}), g^R(\underline{r}_0)$	Probability distribution over initial position r_0 of protons
$G^{i,f}_{\ell}$	Transition matrix elements
G_{γ}, G_{Γ}	Phase factor
g^2_A, g^2_E	Powder average of $\lvert G_A\rvert^2$ and $\lvert G_E\rvert^2$
g	Coupling constant
Γ_{ℓ}, Γ	Relaxation rate
Γ	Symmetry label
γ	Grüneisen constant
$\hbar = h/2\pi$	Planck's constant
$H^{(\ell)}_{mm'}(\tau)$	Rotator functions expressed in terms of quaternions
$\mathcal{H}$	Hamiltonian
$\widehat{\mathcal{H}}, \widehat{\mathcal{H}}_T$	Hamiltonian matrix
h, h_i	120° overlap matrix elements
H, H_x, H_y, H_z	180° overlap matrix elements
$\mathring{i}, \mathring{i}_z, \mathring{i}_+, \mathring{i}_-$	Neutron scattering operators (total, z component, raising and lowering operators), acting on nuclear spin of scatterer
I, I_z	Nuclear spin of scattering nucleus: total spin and z component
$I_S(\underline{Q},t)$	Intermediate scattering function
$I^T_S, I^R_S(\underline{Q},t)$	Translational, rotational intermediate scattering function
$I_1(\underline{Q},t)$	Intermediate one-phonon scattering function
I_{FW}	Intensity of fixed-window measurement
$I(\omega)$	Spectral function
$j_{\ell}(x)$	Spherical Bessel function of order ℓ
$J_{\ell}(x)$	Bessel function of order ℓ
J	Rotational quantum number
$K_{\ell m}$	Cubic harmonics
$\underline{k}, \underline{k}'$	Wave vector of the neutron
K	Kinetic energy
κ	Compressibility
λ	Eigenvalue
M, M_p	Number of atoms (protons) in molecule
M'	Number of sites accessible via jumps

m_N	Neutron mass
M_J	Multiplicity of state with rotational quantum number J
$\bar{\bar{M}}, \bar{\bar{M}}_N$	Matrix of transition matrix elements
$\|\mu\rangle$	Spin function of neutron
$\|\mu_1\mu_2\mu_3\mu_4\rangle$	Spin function of tetrahedral molecule; μ_i denotes spin of particle i
$[\mu_1\mu_2\mu_3\mu_4\rangle$	Spin function of tetrahedral molecule; μ_i denotes spin at position i
N	Number of molecules (sometimes: atoms) in the crystal
n_R	Quantum number of rotational oscillator
n_T	Quantum number of translational oscillator
ω^E	Euler angles (sometimes collectively denotes angles)
$\Omega, \Omega_Q, \Omega^M$	Polar coordinates
$\omega, \omega_{aa'}$	Frequency
ω_0	Ground-state splitting
ω_p	Peak frequency
ω_i	Splitting of ith excited librational state
ω_D	Debye frequency
$P(\{\omega_j^E;\underline{R}_j\}, \{\omega_{j'}^E;\underline{R}_{j'}\}t)$	Density matrix
P_μ, P_a	Statistical weights of quantum-mechanical states
$P_j^{(i)}(t), P_j^{(t)}$	Probability of finding a particle at site $\underline{r}_j$ at time t provided that it was $\underline{r}_{i0}$ at t = 0
P^Γ	Projection operator
ψ, ψ_a	Total wave function of a molecule
$\underline{Q}, Q_\parallel, Q_\perp$	Momentum transfer of the neutrons
$\underline{g}$	Eigenvector in reorientation problems
$\underline{r}$	Positional vector (of atom within molecule)
$\underline{R}_i$	Centre-of-mass coordinate of ith molecule
$\underline{\rho}$	Vector connecting atom with molecular centre
ρ	Radius of molecule (distance from molecular c.o.m.)
$\rho(\underline{r})$	Density or charge distributions
$\hat{R}, \hat{R}, \hat{R}_0$	Rotational operator (symmetry operation)
R	Rotation matrix
$\bar{\bar{R}}^N$	Matrix which diagonalizes Hamiltonian matrix $\mathcal{H}_T$
$\mathring{s}, \mathring{s}_z, \mathring{s}_+, \mathring{s}_-$	Scattering operator acting on neutron (total, z component, raising, lowering)
$S(\underline{Q},\omega)$	Scattering function (van Hove)
$S_s(\underline{Q},\omega)$	Symmetric scattering function
$S_{inc}(\underline{Q},\omega)$	Incoherent scattering function
σ_{inc}	Incoherent scattering cross-section
SAF	Symmetry-adapted function
se_i	Mathieu functions (odd)
t	Time
τ	Residence time

T_1	Spin-lattice relaxation time
τ_R	Characteristic time in Diffusion
T	Temperature
T_0	Transition temperature
Θ_D	Debye temperature
Θ	Moment of inertia of a molecule
θ	Azimuthal angle
$\tau = \tau_1, \tau_2, \tau_3, \tau_4$	Quaternions (coordinates on four-dimensional unit sphere)
$U^{(\ell)}_{mm'}(\omega^E)$	Rotator functions
$\langle u^2 \rangle$	Mean squared amplitude of translation oscillations
u	Probability of 120° jump
$\underline{u}_R$	Rotational displacement
$\underline{u}_T$	Translational displacement
$V_1^{ij}, V_2^{ij}, V, V_n$	Potential
$V_c(\omega^E)$	Crystal field
$V^i(\omega^E, t)$	Single-particle potential
$V^i_{st}(\omega^E_i)$	Static part of single-particle potential
$V^i_{fl}(\omega^E_i, t)$	Fluctuating part of single-particle potential
$V' = V/B$	Scaled potential
V_h	Harmonic part of potential
$v(r)$	Atom-atom potential
v_1, v_2, v	Probability of 90° jump
W, W^Γ	Neutron scattering operator
W^i	Single-particle potential
$W(\alpha)$	Distribution function of angular steps
$w = \tau^{-1}$	Relaxation rate
w_0	Rate of transitions from ground-state (CH_3 group)
w_1	Rate of transitions from first excited state (CH_3 group)
$\bar{\bar{W}}$	Matrix of relaxation rates
$\underline{x}_R$	Rotational axis
$\underline{X}, \underline{X}_c, \underline{X}_M$	Coordinate systems
x_G	Width parameter of pocket state
x	Concentration
ξ, η, ζ	Euler angles
$Y_{\ell m}(\theta, \phi)$	Spherical harmonics
ζ_0	Friction constant

References

1.1 K. Clusius: Z. Phys. Chem. Leipzig *3*, 59 (1929)
1.2 J.N. Sherwood (ed.): *The plastically Crystalline State (Orientationally Disordered Crystals)* (Wiley, Chichester 1979)
1.3 N.E. Parsonage, L.A.K. Staveley: *Disorder in Crystals* (Clarendon, Oxford 1978)
1.4 G. Venkataraman, V.C. Sahni: Rev. Mod. Phys. *42*, 409 (1970)
1.5 G. Dolling: Proceedings of the Conference on Neutron Scattering, Gatlinburg, USA, June 6-10, 1976, ed. by R.M. Moon, Vol. I, p. 263
1.6 G.E. Bacon: *Neutron Diffraction* (Clarendon, Oxford 1975)
1.7 J. White, T. Springer: Phys. Bl. *35*, 398 and 448 (1979), and references therein
1.8 T. Springer: *Quasielastic Neutron Scattering for the Investigation of Diffusive Motions in Solids and Liquids*, Springer Tracts in Modern Physics, Vol. 64 (Springer, Berlin, Heidelberg, New York 1972)
1.9 T. Springer:*"Molecular Rotations and Diffusion in Solids, in Particular Hydrogen in Metals", Dynamics of Solids and Liquids by Neutron Scattering*, ed. by S.W. Lovesey, T. Springer, Topics in Current Physics, Vol. 3 (Springer, Berlin, Heidelberg, New York 1977)
1.10 A.J. Leadbetter, R.E. Lechner: *"Neutron Scattering Studies"*, in Ref. 1.2, p. 285
1.11 A. Hüller, W. Press: *"Rotational Excitations and Tunneling"*, in *Neutron Inelastic Scattering*, Proc. Symposium 1977 (IAEA Vienna 1978) Vol. I, p. 231

2.1 A. Hüller: Z. Phys. *250*, 456 (1972)
2.2 A. Hüller: Z. Phys. *270*, 343 (1974)
2.3 Y. Yamada, M. Mori, Y. Noda: J. Phys. Soc. Jpn. *32*, 1565 (1972)
2.4 H. Yasuda, T. Yamamoto: Prog. Theor. Phys. *45*, 1458 (1971)
2.5 H. Yasuda: Prog. Theor. Phys. *45*, 1361 (1971)
2.6 J.C. Raich, H.M. James: Phys. Rev. Lett. *16*, 173 (1966)
2.7 H.M. James, T.A. Keenan: J. Chem. Phys. *31*, 12 (1959)
2.8 K.H. Michel, J. Naudts: J. Chem. Phys. *67*, 547 (1977)
2.9 L.D. Landau, E.M. Lifschiz: *Course of Theoretical Physics, Vol. 3, Quantum Mechanics* (Pergamon, New York 1964)
2.10 A. Hüller, W. Press: *"Theoretical Aspects of Solid Rotator Phases"*, in Ref. 1.2, p. 345
2.11 I.F. Silvera: Rev. Mod. Phys. *52*, 393 (1980)
2.12 W. Press: J. Chem. Phys. *56*, 2597 (1972)
2.13 D.E. Cox, E.J. Samuelsen, K.H. Beckurts: Phys. Rev. B *7*, 3102 (1973)
2.14 F.C. von der Lage, H.A. Bethe: Phys. Rev. *71*, 61 (1947)
2.15 S.L. Altmann, A.P. Cracknell: Rev. Mod. Phys. *37*, 19 (1965)
2.16 A. Hüller, J. Kane: J. Chem. Phys. *61*, 3599 (1974)
2.17 A. Hüller, W. Press: Acta Crystallogr. A *35*, 876 (1979)
2.18 A.I. Kitaigorodsky: *Molecular Crystals and Molecules* (Academic, New York 1973)
2.19 T. Kihara: *Intermolecular Forces* (Wiley, New York 1977), and references therein
2.20 T.B. McRury, W.A. Steele: J. Chem. Phys. *64*, 1288 (1976)

2.21 B.C. Carlson, G.S. Rushbrooke: Proc. Cambridge Philos. Soc. *46*, 626 (1950)
2.22 T. Yamamoto, Y. Kataoka, K. Okada: J. Chem. Phys. *66*, 2701 (1977)
2.23 J. Bartholomé, R. Navarro, D. Gonzâlez, L.J. deJongh: Physica B *92*, 22 (1977)
2.24 W. Press, A. Hüller: In *Anharmonic Lattices, Structural Transitions and Melting*, ed. by T. Riste (Noordhoff, Leiden 1974)
2.25 J. Eckert, W. Press: J. Chem. Phys. *73*, 451 (1980)
2.26 P. Dunmore: J. Low Temp. Phys. *24*, 397 (1976)
2.27 W. Press, A. Hüller: J. Chem. Phys. *68*, 4465 (1978)
2.28 J.M. Rowe, D.G. Hinks, D.L. Price, S. Susman, J.J. Rush: J. Chem. Phys. *58*, 2039 (1973)
2.29 A.F. Devonshire: Proc. R. Soc. London A *153*, 601 (1936)
2.30 F. Hund: Z. Phys. *51*, 1 (1928)

3.1 W. Marshall, S.W. Lovesey: *Theory of Thermal Neutron Scattering* (Clarendon, Oxford 1971)
3.2 L. van Hove: Phys. Rev. *95*, 249 (1954)
3.3 A. Hüller: Phys. Rev. B *16*, 1844 (1977)
3.4 A. Hüller, W. Press: Phys. Rev. B (1981)
3.5 S.K. Sinha, G. Venkataraman: Phys. Rev. *149*, 1 (1966)
3.6 J. Hama, H. Miyagi: Prog. Theor. Phys. *50*, 1142 (1973)
3.7 W. Press, H. Grimm, A. Hüller: Acta Crystallogr. A *35*, 881 (1979)
3.8 V. Lottner, A. Heim, T. Springer: Z. Phys. B *32*, 157 (1979)
3.9 V.F. Sears: Can. J. Phys. *44*, 1299 (1966)
3.10 V.F. Sears: Can. J. Phys. *45*, 237 (1967)
3.11 J.R.D. Copley: Comput. Phys. Commun. *7*, 289 (1974)
3.12 A. Bickermann: *"Inkohärente Neutronenstreuung am festen Wasserstoff"*; Tech. Rpt. JÜL-1487, Institut für Festkörperforschung, Kernforschungsanlage Jülich (1978)

4.1 F. Reif: *Fundamentals of Statistical and Thermal Physics* (McGraw-Hill, New York 1965)
4.2 P.A. Egelstaff, P. Schofield: Nucl. Sci. Eng. *12*, 260 (1962)
4.3 D. Richter, B. Ewen: In *Neutron Inelastic Scattering*, Proc. Symp. 1977, (IAEA, Vienna 1978) Vol. I, p. 589
4.4 J.A. Janik, J.M. Janik, K. Otnes, K. Rosciszewski: Physica B *83*, 259 (1976)
4.5 P.A. Egelstaff: J. Chem. Phys. *53*, 2590 (1970)
4.6 E.N. Ivanov: Sov. Phys. JETP *18*, 1041 (1964)
4.7 V.F. Sears: Can. J. Phys. *44*, 1279 (1966)
4.8 B. de Raedt: Thesis, Saarbrücken 1979
4.9 K.E. Larson, T. Mansson, L.E. Olsson: In *Neutron Inelastic Scattering*, Proc. Symp. 1977 (IAEA, Vienna 1978) Vol. I, p. 435
4.10 R.E.W. Wyckoff: *Crystal Structures* (Wiley-Interscience, New York 1969) Vols. II, III
4.11 W. Pieszek, G.R. Strobl, K. Malzahn: Acta Crystallogr. B *30*, 1278 (1974)
4.12 K. Kurki-Suonio: Ann. Acad. Sci. Fenn. Ser. A *4*, 241 (1967)
4.13 R.S. Seymour, A.W. Pryor: Acta Crystallogr. B *26*, 1487 (1970)
4.14 W. Press, A. Hüller: Acta Crystallogr. A *29*, 252 (1973)
4.15 J.D. Barnes: J. Chem. Phys. *58*, 5193 (1973)
4.16 G.G. Windsor, D. Bloor, D.H. Bonsor, D.N. Batchelder: Internal Rpt. MPD/NBS/38, AERE, Harwell (1977)
4.17 B. Ewen, D. Richter: J. Chem. Phys. *69*, 2954 (1978)
4.18 M.L. Klein, J.J. Weis: J. Chem. Phys. *67*, 217 (1977)
4.19 W. Press, U. Buchenau: unpublished
4.20 R. Stockmeyer, H. Stiller: Phys. Status Solidi *27*, 269 (1968)
4.21 K. Sköld: J. Chem. Phys. *49*, 2443 (1968)
4.22 C. Thibaudier, F. Volino: Mol. Phys. *26*, 1281 (1973)
4.23 C. Thibaudier, F. Volino: Mol. Phys. *30*, 1159 (1975)
4.24 P. Rigny: Physica *59*, 707 (1972)
4.25 C. Brot, B. Lassier-Govers: Ber. Bunsenges. Phys. Chem. *80*, 31 (1976)

4.26 K. Roscieszewski: unpublished
4.27 M. Prager, W. Press, B. Alefeld, A. Hüller: J. Chem. Phys. *67*, 5126 (1977)
4.28 J. Töpler, D.R. Richter, T. Springer: J. Chem. Phys. *69*, 3170 (1978)
4.29 J.A. Lersbscher, J. Trotter: Acta Crystallogr. B *32*, 2671 (1976)
4.30 E.O. Schlemper, W.C. Hamilton, J.J. Rush: J. Chem. Phys. *44*, 2499 (1966)
4.31 K. Otnes, I. Svare: J. Phys. C *12*, 3899 (1979)
4.32 A. Kollmar, B. Alefeld: Proc. of the Conf. on Neutron Scattering, Gatlinburg, USA, June 6-10, 1976, ed. by R.M. Moon, Vol. I, p. 330
4.33 H.A. Levy, S.W. Peterson: Phys. Rev. *86*, 766 (1952)
4.34 K.H. Michel: J. Chem. Phys. *58*, 1143 (1973)
4.35 L. Katz, M. Guinan, R.J. Borg: Phys. Rev. B *4*, 330 (1971)
4.36 G. Venkataraman, K. Usha, P.K. Iyengar, R.P. Vijayaraghavan, A.P. Roy: In *Inelastic Scattering in Solids and Liquids*, Proc. Symp. Chalk River 1962 (IAEA, Vienna 1963) Vol. II, p. 253
4.37 R. Gerling, A. Hüller: Z. Phys. B *40*, 209 (1980)
4.38 A.J. Dianoux, F. Volino: Mol. Phys. *34*, 1263 (1977)
4.39 P. Fulde, L. Pietronero, W.R. Schneider, S. Strässler: Phys. Rev. Lett *35*, 1776 (1975)
4.40 T. Geisel: "Continuous Stochastic Models", in *Physics of Superionic Conductors*, ed. by M.B. Salomon, Topics in Current Physics, Vol. 15 (Springer, Berlin, Heidelberg, New York 1979) Chap. 9
4.41 B. De Raedt, K.H. Michel: Phys. Rev. B *19*, 767 (1979)
4.42 B. de Raedt, K.H. Michel: Faraday Discuss. Chem. Soc. *69*, 88 (1980)
4.43 R. Callender, P.S. Pershan: Phys. Rev. A *2*, 672 (1970)

5.1 P.A. Egelstaff, B.C. Haywood, F.J. Webb: Proc. Phys. Soc. London *90*, 681 (1967)
5.2 H. Stein, H. Stiller, R. Stockmeyer: J. Chem. Phys. *57*, 1726 (1972)
5.3 H. Kapulla, W. Gläser: Phys. Lett. A *31*, 158 (1970)
5.4 H. Kapulla, W. Gläser: In *Neutron Inelastic Scattering*, Proc. Symp. Grenoble, France, 1972 (IAEA, Vienna 1973) p. 841
5.5 W. Press, A. Kollmar: Solid State Commun. *17*, 405 (1975)
5.6 M. Abramowitz, I.A. Segun: *Handbook of Mathematical Functions* (Dover, New York 1968)
5.7 N.W. McLachlan: *Theory and Application of the Mathieu Function* (Clarendon, Oxford 1947)
5.8 R.F. Gloden: *Euratom* Rpt. EUR 4349f (1970); EUR 4348 (1970)
5.9 H.F. King, D.F. Hornig: J. Chem. Phys. *44*, 4520 (1966)
5.10 S.C. Jain, V.K. Tewary: J. Phys. C *6*, 1999 (1973)
5.11 Manashi Roy, C.K. Majumdar: Solid State Commun. *14*, 695 (1974)
5.12 D. Smith: J. Chem. Phys. *58*, 3833 (1973)
5.13 A. Hüller, D.M. Kroll: J. Chem. Phys. *63*, 4495 (1975)
5.14 A. Hüller, J. Raich: J. Chem. Phys. *71*, 3851 (1979)
5.15 Y. Kataoka, W. Press, U. Buchenau, H. Spitzer: In *Neutron Inelastic Scattering*, Proc. Symp. 1977 (IAEA, Vienna 1978), Vol. I, p. 311
5.16 V. Narayanamurti, R.O. Pohl: Rev. Mod. Phys. *42*, 201 (1970)
5.17 A.S. Barker, Jr., A.J. Sievers: Rev. Mod. Phys. *47*, 2 (1975)
5.18 D. Walton, H.A. Mook, R.M. Nicklow: Phys. Rev. Lett. *33*, 412 (1974)
5.19 M. Mostoller, R.F. Wood: Proc. of the Conf. on Neutron Scattering, Gatlinburg, USA, June 6-10, 1976, ed. by R.M. Moon, Vol. I, p. 167
5.20 B. Janik, U. Buchenau, W. Press: unpublished
5.21 D. Smith: J. Chem. Phys. *68*, 3222 (1978)
5.22 P. Sauer: Z. Phys. *194*, 360 (1972)
5.23 H.U. Beyeler: Phys. Status Solidi B *52*, 419 (1972)
5.24 H.U. Beyeler: Phys. Rev. B *11*, 3078 (1975)
5.25 T. Nagamiya: Prog. Theor. Phys. *6*, 702 (1951)
5.26 D. Smith: J. Chem. Phys. *66*, 4587 (1977)
5.27 D. Smith: J. Chem. Phys. *68*, 619 (1978)
5.28 D. Smith: to be published

5.29 K. Nishiyama, T. Yamamoto: J. Chem. Phys. *58*, 1001 (1973)
5.30 K. Kobashi, Y. Kataoka, T. Yamamoto: Can. J. Chem. *54*, 2154 (1976)
5.31 Y. Kataoka, K. Okada, T. Yamamoto: Chem. Phys. Lett. *19*, 365 (1973)
5.32 R.F. Code, J. Higinbotham: Can. J. Phys. *54*, 1248 (1976)
5.33 A.J. Nijman, A.J. Berlinsky: Phys. Rev. Lett. *38*, 408 (1977)
5.34 C.H. Anderson, N.F. Ramsey: Phys. Rev. *149*, 14 (1966)
5.35 Y. Ozaki, Y. Kataoka, T. Yamomoto: J. Chem. Phys. *73*, 3442 (1980)
5.36 A.C. Zemach, R.J. Glauber: Phys. Rev. *101*, 118 (1956)
5.37 G.W. Griffing: Phys. Rev. *124*, 1489 (1961)
5.38 J.A. Janik, A. Kowalska: In *Thermal Neutron Scattering*, ed. by P.A. Egelstaff (Academic, New York 1965)
5.39 W. Press, M. Prager, A. Heidemann: J. Chem. Phys. *72*, 5924 (1980)
5.40 G. Sarma: J. Phys. Radium *21*, 783 (1960)
5.41 Y.D. Harker, R.M. Brugger: J. Chem. Phys. *66*, 2701 (1977)
5.42 A. Hüller, M. Prager: Solid State Commun. *29*, 537 (1979)

6.1 W. Schott: Z. Phys. *231*, 243 (1970)
6.2 M. Nielsen: Phys. Rev. B *7*, 1626 (1973)
6.3 J.L. Yarnell, R.L. Mills, A.F. Schuch: Int. Conf. on Quantum Crystals, Tbilissi, USSR (1974)
6.4 W. Press: Acta Crystallogr. A *29*, 257 (1973)
6.5 A.B. Harris, A.J. Berlinsky: Phys. Rev. B *16*, 3791 (1977)
6.6 W.N. Hardy, A.J. Berlinsky, A.B. Harris: Can. J. Phys. *55*, 1150 (1977)
6.7 G.W. Griffing: In *Inelastic Scattering of Neutrons in Solids and Liquids*, Proc. Symp. Chalk River 1962 (IAEA, Vienna 1963) p. 435
6.8 A.J. Nijman: "Investigation of Solid Methane by NMR at High Pressure and Low Temperature"; Thesis, Univ. Amsterdam (1977)
6.9 K. Maki, Y. Kataoka, T. Yamamoto: J. Chem. Phys. *70*, 655 (1979)
6.10 J. Eckert, J.A. Goldstone, C.R. Fincher, jr., W. Press: Bull. Am. Phys. Soc. (1980) and to be published in J. Chem. Phys.
6.11 A. Hüller: Z. Phys. B *36*, 25 (1980)
6.12 A.E. Zweers, H.B. Brom, W.J. Huiskamp: Phys. Lett. A *47*, 347 (1974)
6.13 B. Alefeld, A. Kollmar, B.A. Dasannacharya: J. Chem. Phys. *63*, 4415 (1975)
6.14 H. van Kempen, T. Garofano, A.R. Miedema, W.J. Huiskamp: Physica *31*, 1096 (1965)
6.15 J.H. Colwell, E.K. Gill, J.A. Morrison: J. Chem. Phys. *42*, 3144 (1965)
6.16 J. Haupt, W. Müller-Warmuth: Z. Naturforsch *23a*, 208 (1968)
6.17 H. Glättli, A. Sentz, M. Eisenkremer: Phys. Rev. Lett. *28*, 871 (1972)
6.18 M. Punkkinen: J. Magn. Reson. *19*, 222 (1975)
6.19 M. Punkkinen, J.E. Tuohi, E.E. Ylinen: Chem. Phys. Lett. *36*, 393 (1975)
6.20 R.S. Hallsworth, D.W. Nicoll, J. Peternelj, M.M. Pintar: Phys. Rev. Lett. *39*, 1493 (1977)
6.21 B. Alefeld, A. Kollmar: Phys. Lett. A *57*, 289 (1976)
6.22 M.G. Miksic, E. Segerman, B. Post: Acta Crystallogr. *12*, 390 (1959)
6.23 P.S. Allen: J. Phys. C *7*, L22 (1974)
6.24 S. Clough, S.M. Nugent: J. Phys. C *9*, L561 (1976)
6.25 S. Clough, A. Heidemann: In *Neutron Inelastic Scattering*, Proc. Symp. 1977 (IAEA, Vienna 1978) Vol. I, p. 255
6.26 S. Clough, A. Heidemann: J. Phys. C *12*, 761 (1979)
6.27 S. Clough, A. Heidemann, K. Kraxenberger: Phys. Rev. Lett. *42*, 1298 (1979)
6.28 S. Clough, B. Alefeld, J.B. Suck: Phys. Rev. Lett. *41*, 124 (1978)
6.29 P.A. Beckmann, S. Clough: J. Phys. C *10*, L231 (1977)
6.30 W. Müller-Warmuth, R. Schüler, A. Kollmar, M. Prager: In *Magnetic Resonance and Related Phenomena*, ed. by H. Brunner, K.H. Hauser, D. Schweitzer; Proc. of the XIXth Congress Ampère, Heidelberg, 1976 (Groupement Ampère, Heidelberg 1976) p. 345
6.31 M. Prager, A. Kollmar, W. Müller-Warmuth, R. Schüler: In *Neutron Inelastic Scattering*, Proc. Symp. 1977 (IAEA, Vienna 1978) Vol. I, p. 265

6.32 W. Müller-Warmuth, R. Schüler, M. Prager, A. Kollmar: J. Chem. Phys. *69*, 2382 (1978)
6.33 W. Müller-Warmuth, R. Schüler, M. Prager, A. Kollmar: J. Magn. Reson. *34*, 83 (1979)
6.34 J. Haupt: Z. Naturforsch. *26a*, 1578 (1971)
6.35 L.A. de Graaf, C. Steenbergen: Physica B *97*, 199 (1979)
6.36 B. Alefeld, W. Press: to be published
6.37 A.R. Bates, K.W.H. Stevens: J. Phys. C *2*, 1573 (1969)
6.38 W. Press, M. Prager: J. Chem. Phys. *67*, 5752 (1977)
6.39 W. Press, A. Hüller: Phys. Rev. Lett. *30*, 1207 (1973)
6.40 A. Heidemann, W. Press, K. Lushington, J. Morrison: to be published
6.41 M.W. Newbery, T. Rayment, M.V. Smalley, R.K. Thomas, J.W. White: Chem. Phys. Lett. *59*, 461 (1978)
6.42 G. Bomchil, A. Hüller, T. Rayment, S.J. Roser, M.V. Smalley, R.K. Thomas, J.W. White: Phil. Trans. R. Soc. London B *290*, 537 (1980)
6.43 M. Punkkinen, E.E. Ylinen, L.P. Ingman: "Librational Energy Levels of the NH_4^+ Ions in $(NH_4)_2SnCl_6$ by NMR", in *Magnetic Resonance and Related Phenomena*, Proc. XXth Congress Ampère, Tallinn, USSR, August 21-26, 1978, ed. by E. Kundla, E. Lippmaa, T. Saluvere (Springer, Berlin, Heidelberg, New York 1980) p. 139
6.44 M. Prager, B. Alefeld: J. Chem. Phys. *65*, 4927 (1976)
6.45 M. Prager, B. Alefeld, A. Heidemann: In *Magnetic Resonance and Related Phenomena*, ed. by H. Brunner, K.H. Hauser, D. Schweitzer, XIXth Congress Ampère, Heidelberg 1976, p. 389 (Groupement Ampère, Heidelberg 1976)
6.46 C.S. Choi, H.J. Prask, E. Prince: J. Chem. Phys. *61*, 3523 (1974)
6.47 W. Güttler, J.U. v. Schütz: Chem. Phys. Lett. *20*, 133 (1973)
6.48 M. Punkkinen, J.P. Pyy: Phys. Fenn. *10*, 215 (1975)
6.49 M. Prager, W. Press: J. Chem. Phys. *75*, 492 (1981)

7.1 M. Punkkinen: Phys. Rev. B *21*, 54 (1980)
7.2 M. Punkkinen: J. Phys. C *11*, 3039 (1978)
7.3 P.W. Anderson: J. Phys. Soc. Jpn. *9*, 316 (1954)
7.4 A. Abragam: *Principles of Nuclear Magnetism*, Oxford U.P., (London 1961) p. 447
7.5 I. Svare: J. Phys. C *12*, 3907 (1979)
7.6 S. Clough, J.R. Hill: J. Phys. C *9*, L645 (1976)
7.7 T. Holstein: Ann. Phys. *8*, 325 and 343 (1959)
7.8 T. Moriya: Prog. Theor. Phys. *18*, 567 (1957)
7.9 G. Briganti, P. Calvani, F. Deluca, B. Maraviglia: Can. J. Phys. *56*, 1182 (1978)
7.10 M. Prager, W. Press: unpublished
7.11 S. Clough, A. Heidemann, M. Paley, C. Vettier: J. Phys. C *12*, L781 (1979)
7.12 K. Kobashi, T. Kihara: J. Chem. Phys. *72*, 378 (1980)
7.13 P.P. Peressini, J.P. Harrison, R.O. Pohl: Phys. Rev. *180*, 926 (1969)
7.14 K.J. Lushington, J.A. Morrison: J. Chem. Phys. *69*, 4214 (1978)
7.15 M.A. White, K.J. Lushington, J.A. Morrison: J. Chem. Phys. *69*, 4227 (1978)
7.16 M.A. White, J.A. Morrison: J. Chem. Phys. *70*, 5384 (1979)
7.17 M. Prager, W. Press, A. Heidemann: to be published in J. Chem. Phys.
7.18 G. Arzi, G. Sandor: Acta Crystallogr. A *31*, 188 (1975)
7.19 K. Maki, Y. Kataoka, T. Yamamoto: J. Chem. Phys. *70*, 655 (1979)
7.20 N.S. Sullivan, M. Devoret, B.P. Cowan, C. Urbina: Phys. Rev. B *17*, 5016 (1978)
7.21 J.M. Rowe, J.J. Rush, D.G. Hinks, S. Suman: Phys. Rev. Lett. *43*, 1158 (1979)
7.22 C.S. Barrett, L. Meyer: J. Chem. Phys. *42*, 107 (1965)
7.23 M. Prager, W. Press, K. Rössler: J. Molec. Spectrosc. *60*, 173 (1980)
7.24 W. Press, U. Buchenau, A. Kollmar, F. Maniawski: to be published
7.25 E. Eucken, H. Veith: Z. Phys. Chem. B *24*, 275 (1936)
7.26 W. Press, B. Janik: to be published
7.27 R.C. Zeller, R.O. Pohl: Phys. Rev. B *4*, 2029 (1971)
7.28 H. Boysen, A.W. Hewat: Acta Crystallogr. B *34*, 1412 (1978)

Subject Index

Errata

The following corrections were received after a portion of the book had been printed.

page 9 Table 2.1 *please read*

Static potential V'_{st} \ Fluctuating potential V'_{fl}

page 64 line 7 from the top *please read* ... symmetrized functions ...

page 94 one line above Eq.(7.1) *please read* ... for the spectral function ...

page 102 *please move the last sentence of the 2nd paragraph* ... The application of high ... *to the end of the first paragraph* ... exponent n. The application of high ...

page 106 *unnumbered equation* ... $= \sum_j V(\omega_i^E, \omega_j^E)$

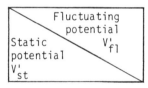